COVID-19

A Complex Systems Approach

Papers and Commentaries

Editors: Alfredo J. Morales, Joseph Norman
and Mads Bahrami.

STEM
ACADEMIC PRESS

STEM Academic Press operates under an academic journal-style board and publishes books containing peer-reviewed material in the mathematical and quantitative sciences. A colored electronic version of this volume is freely available to the general public at http://www.stemacademicpress.com/.

Covid-19: A Complex Systems Approach
STEM Volumes
Keywords: Complex Systems | Epidemics | Modeling | Data Analytics
ISBN 978-0-578-91200-4
STEM Academic Press, New York, USA
© 2021 Mads Bahrami

Contents

Introduction

The Editors

We are all connected. Following the year of 2020, everyone must recognize this not as an abstract statement or mere platitude, but a concrete reality with its potential catastrophic risks that we are obligated to grapple with.

While our ability to connect with one another offers many opportunities for collaboration and cooperation, it also brings with it risks of cascade from multiplicative processes that propagate via these connections. With the COVID-19 pandemic, everyone now has a clear firsthand experience of such complexity and its unimaginable consequences: how remote events in far regions can endanger one's survival, literally, in a matter of few months.

The COVID-19 pandemic, which continues into 2021 with no certain bound, has unfortunately demonstrated what many have been warning about for years: we are vulnerable to global contagion. A very important question is why? Given all advances in sciences and technologies, our unprecedented computational ability with supercomputers, why almost all countries in the world failed badly in dealing with the COVID-19 outbreak? What went wrong?

We argue that due to nonlinear scaling behavior in complex systems (say, our connected world today), we need a fundamental shift in paradigm, tactics and strategy when it comes to treating diseases at the collective scale. Individual medicine does not simply 'scale up' to the collective, and neither does collective mitigation necessarily 'scale down' to the individual. For example, rapid tests that are viewed as too unreliable as an individual diagnostic instrument, could be sufficiently reliable at the collective scale when deployed widely.

In general, multi-scale problems demand multi-scale solutions. The macroscopic behaviors must be addressed first in order to gain a foothold from which one can pursue more granular approaches. This includes, for instance, pausing or imposing mitigation like testing or quarantine for travel between regions when many locales may be spared or otherwise have much more tractable problems when not seeded from external arrivals. This also includes identifying the statistical profile of the collective that allows a pathogen to persist and amplify in a population.

The COVID-19 pandemic shows that the tail of the distribution – the so-called "super-spreader" events – can be enough to sustain an epidemic even when most individuals that become infected are able to avoid infecting others. In other words, it is possible, within some range of parameters, to induce decay of an epidemic simply by removing the extreme spreading events, effectively reducing the replication rate R_0 to below 1.

Early in the COVID-19 outbreak, many precautionary measures were dismissed in reaction to their perceived costliness. The overwhelming costs we have witnessed since, in terms of human life and well-being, societal tension and unrest, and economic hardship should be testament enough to the relative cheapness of paying for insurance up front to avoid paying much more over a much longer time.

In this volume we have included a variety of studies from varying perspectives. The unifying theme in these studies is that the COVID-19 pandemic (and all epidemics and potential pandemics) are essentially systems problems, and must be studied and combated as such. In this selection of papers on COVID-19, the problem has been investigated using a complex systems approach, including agent-based models, cellular automata, networks, population dynamics, spatial-temporal patterns, risk management, analysis of fat-tails, data analysis and visualization.

We can categorize this collection into the following parts:

- *Perspectives*: papers discuss policies and strategies against an outbreak, especially when nothing is clear enough yet, and there are lots of unknown and uncertain factors involved.

- *Probabilities*: papers elaborate on the importance of proper understanding of probability and its role in decision-making against a contagion outbreak.

- *Modeling*: papers explain different approaches to model the spread of a virus in a population.

- *Data*: the main focus of these papers is on data, data collecting, wrangling and manipulation, and its relevance in a fight against a contagion outbreak.

- *Policy*: these papers elaborate on effective policies and strategies one can adapt for future.

We believe this collection can help scientists and decision-makers alike in the direction of a paradigm shift in tactics and strategies when dealing with a pandemic. Last but not least, despite the destructiveness of COVID-19 pandemic, we can't lose track of the fact that until we learn how to grapple with contagion in our connected world, we will be at risk of much more damaging events.

Systemic Risk of Pandemic via Novel Pathogens – Coronavirus: A Note

Joseph Norman[1,2,*], Yaneer Bar-Yam[1] and Nassim Nicholas Taleb[3,4]

[1]New England Complex Systems Institute
[2]Applied Complexity Science, LLC
[3]School of Engineering, New York University
[4]Universa Investments
[*]Corresponding author: joe@appliedcomplexitysciences.com

Note added after original publication: This short note was originally published online on January 26th, 2020. At that time the COVID-19 outbreak was not considered a pandemic (how widespread it was then remains an open question), and authorities such as the World Health Organization were actively discouraging any lifestyle modifications that would mitigate the risk of spreading the disease. The note focuses primarily on exercising precaution via macroscopic mobility constraints given how little was known at the time. It remains that taking these actions would have produced a much better posture from which to attack the COVID-19 outbreak, and nothing in the interim has altered that conclusion.

1   Introduction

The novel coronavirus emerging out of Wuhan, China has been identified as a deadly strain that is also highly contagious. The response by China to date has included travel restrictions on tens of millions across several major cities in an effort to slow its spread. Despite this, positively identified cases have already been detected in many countries spanning the globe and there are doubts such containment would be effective. This note outlines some principles to bear in relation to such a process.

Clearly, we are dealing with an extreme fat-tailed process owing to an increased connectivity, which increases the spreading in a nonlinear way [1,2]. Fat tailed processes have special attributes, making conventional risk-management approaches inadequate.

2   General Precautionary Principle

The general (non-naive) precautionary principle [3] delineates conditions where actions must be taken to reduce risk of ruin, and traditional cost-benefit analyses must not be used. These are ruin problems where, over time, exposure to tail events leads to a certain eventual extinction. While there is a very high probability for humanity surviving a single such event, over time, there is eventually zero probability of surviving repeated exposures to such events. While repeated risks can be taken by individuals with a limited life expectancy, ruin exposures must never be taken at the systemic and collective level. In technical terms, the precautionary principle applies when traditional statistical averages are invalid because risks are not ergodic.

3   Naive Empiricism

Next we address the problem of naive empiricism in discussions related to this problem.

Spreading rate

Historically based estimates of spreading rates for pandemics in general, and for the current one in particular, underestimate the rate of spread because of the rapid increases in transportation connectivity over recent years. This means that expectations of the extent of harm are underestimates both because events are inherently fat tailed, and because the tail is becoming fatter as connectivity increases.

Global connectivity is at an all-time high, with China one of the most globally connected societies. Fundamentally, viral contagion events depend on the interaction of agents in physical space, and with the forward-looking uncertainty that novel outbreaks necessarily carry, reducing connectivity temporarily to slow flows of potentially contagious individuals is the only approach that is robust against misestimations in the properties of a virus or other pathogen.

Reproductive ratio

Estimates of the virus's reproductive ratio R_0—the number of cases one case generates on average over the course of its infectious period in an otherwise uninfected population—are biased downwards. This property comes from fat-tailedness [4] due to individual 'superspreader' events. Simply, R_0 is estimated from an average which takes longer to converge as it is itself a fat-tailed variable.

Mortality rate

Mortality and morbidity rates are also downward biased, due to the lag between identified cases, deaths and reporting of those deaths.

Increasingly fatal rapidly spreading emergent pathogens

With increasing transportation we are close to a transition to conditions in which extinction becomes certain both because of rapid spread and because of the selective dominance of increasingly worse pathogens. [5]

Asymmetric uncertainty

Properties of the virus that are uncertain will have substantial impact on whether policies implemented are effective. For instance, whether contagious asymptomatic carriers exist. These uncertainties make it unclear whether measures such as temperature screening at major ports will have the desired impact. Practically all the uncertainty tends to make the problem potentially worse, not better, as these processes are convex to uncertainty.

Fatalism and inaction

Perhaps due to these challenges, a common public health response is fatalistic, accepting what will happen because of a belief that nothing can be done. This response is incorrect as the leverage of correctly selected extraordinary interventions can be very high.

4 Conclusion

Standard individual-scale policy approaches such as isolation, contact tracing and monitoring are rapidly (computationally) overwhelmed in the face of mass infection, and thus also cannot be relied upon to stop a pandemic. Multiscale population approaches including drastically pruning contact networks using collective boundaries and social behavior change, and community self-monitoring, are essential.

Together, these observations lead to the necessity of a precautionary approach to current and potential pandemic outbreaks that must include constraining mobility patterns in the early stages of an outbreak, especially when little is known about the true parameters of the pathogen.

It will cost something to reduce mobility in the short term, but to fail do so will eventually cost everything—if not from this event, then one in the future. Outbreaks are inevitable, but an appropriately precautionary response can mitigate systemic risk to the globe at large. But policy- and decision-makers must act swiftly and avoid the fallacy that to have an appropriate respect for uncertainty in the face of possible irreversible catastrophe amounts to "paranoia," or the converse a belief that nothing can be done.

Bibliography

[1] Y. Bar-Yam, "Dynamics of Complex Systems", 1997.

[2] Y. Bar-Yam, "Transition to extinction: Pandemics in a connected world," 2016.

[3] N.N, Taleb, R. Read, R. Douady, J. Norman, and Y. Bar-Yam, "The precautionary principle (with application to the genetic modification of organisms)," arXiv:1410.5787, 2014.

[4] N.N. Taleb, "The statistical consequences of fat tails,", STEM Academic Press, 2020.

[5] E.M. Rauch and Y. Bar-Yam, "Long-range interactions and evolutionary stability in a predator-prey system," Physical Review E, vol. 73, no. 2, p. 020903, 2006.

Naive Probabilism and Covid-19

Harry Crane*

Rutgers, the State University of New Jersey, Researchers.One
and London Mathematical Laboratory.
www.harrycrane.com
*Corresponding author: hcrane@stat.rutgers.edu

1 Introduction

Consider the guidance from the scientific establishment and government agencies in the early days and weeks of the 2020 U.S. Covid-19 outbreak.

- Masks don't work.
- Travel restrictions are xenophobic.
- No evidence of asymptomatic spread.
- No evidence for human-to-human transmission.
- No evidence that animals can get Covid-19.
- The risk to the general public is low.

We were told these repeatedly by the World Health Organization (WHO), Jerome Adams (U.S. Surgeon General), Anthony Fauci (Director of the U.S. National Institute of Allergy and Infectious Diseases), and a number of other experts in virology, epidemiology and public health. Similar statements downplaying the threat were made by a number of public intellectuals, including Cass Sunstein, John Ioannidis, Richard Epstein, Carl Bergstrom, and several others from academia.

Every one of the above assertions was later disavowed or disproved. In many cases the initially extreme underreaction was matched with an equal overreaction to the opposite extreme. In many jurisdictions, masks were mandated by law in all public places, under penalty of fine. Travel restrictions to, from and within many countries in North America and Europe became the norm. Large gatherings were deemed "superspreader events" because of their tendency for large-scale, human-to-human transmission by asymptomatic participants. Family members were prohibited from visiting hospitalized relatives who were sick and dying from Covid-19. The state of California limited Thanksgiving gatherings to at most 3 households and "strongly discouraged" singing, chanting or shouting. Religious gatherings were severely curtailed and even outlawed in some parts of the

United States, a restriction that was later judged unconstitutional by the U.S. Supreme Court.

The earliest missteps contributed to higher death tolls, prolonged lock-downs, and diminished trust in science and government leadership. The later *post hoc* counter-reactions in the form of indefinite lockdowns, ex-panded government mandates, and extended use of emergency powers contributed further to economic and psychological distress while increasing distrust of the scientific establishment. Yet the organizations and individu-als most responsible for misleading the public and damaging institutional credibility suffered little or no consequences compared to the losses their decisions inflicted on the rest of society. In fact, several of these individuals benefited from their consequential mistakes.

I discuss here how the mistakes listed above result from two related consequences of applying formulaic decision procedures to address highly uncertain and complex problems:

(i) *Naive Probabilism*, the belief that decision-making under uncertainty boils down to probability calculations and statistical analysis; and

(ii) *Freeroll Effect*, the phenomenon by which those responsible for risky policies suffer none of their negative consequences, while benefiting from their upside.

Though both are widespread throughout society, academia, business, medicine, finance, law and politics, I focus here on Naive Probabilism and the Freeroll Effect in U.S. Covid-19 response. In a series of three vignettes, I discuss how Naive Probabilism negatively impacted the U.S. response, and how those responsible for poor policy decisions benefited from the Freeroll Effect, in many cases enjoying increased public profile and greater influence even after negatively impacting the lives of hundreds of millions in the United States and around the world. The "axioms" corresponding to each section give an insight into the thought process underlying Naive Probabilism, and should be interpreted as "anti-axioms" for those wishing to avoid the mistakes highlighted below.

2   A mask is a thing

[First Axiom of Naive Probabilism] *The more complex the problem, the more complicated the solution.*

A hallmark of naive decision making, and especially naive probabilism, is the belief that the more complex the problem, the more sophisticated the solution needs to be. Without question, the Covid-19 outbreak was highly complex due to the uncertainty of its origins, the novelty of the virus, and the interconnectivity of global economies and societies. Of its many complexities, however, the question of masks wasn't one of them. And because of the initial mistakes on masks—in not only failing to recommend but in actively discouraging mask use early on—the U.S. and global agencies (WHO, CDC), its leaders (Fauci), and much of the scientific establishment caused even greater confusion, delayed necessary precautions, and lost public trust to an extent which undermined their credibility on all Covid-19-related matters moving forward.

The mask mistake was consequential not because masks were the antidote to Covid-19—the extent to which masks help remains unknown—but because they were a low cost measure whose effect would be neutral at worst. Expressed in common sense (Northeast Philadelphia) logic: a mask is a thing. Plain and simple. A mask is a thing, a virus is a thing; when one thing gets in the way of another thing, it can prevent that other thing from getting by; therefore, wearing a mask, in worst case, can't hurt in reducing the spread of a virus. This argument is common sense to every plumber (union and non-union), cop, used car salesman, and Uber driver, but not to Drs. Fauci, Adams, and the many other MDs, PhDs and MPHs whose advice influenced early Covid-19 response in the United States.

These experts neglected common sense in favor of a more "scientific response", one based on "rigorous peer review" and sufficient data. Two months after the initial U.S. outbreak, a peer-reviewed study [6] confirmed what common sense knew all along: masks are things. Based on this science, not common sense, masks went from strongly discouraged[1] to mandated by law. Precious time wasted, thousands of lives lost, the economy stalled out, indefinite uncertainty for hundreds of millions around the world, and a drastic reversal from active discouragement of mask-wearing to government decree. First, ill-informed and misleading guidance. Later, over-bearing and disproportionate mandates in response to the initial folly.

As naive, and costly, as the initial mask denial was, it was naive in a way that the "experts" agreed. Even at the time, the mask mistake, which seemed easily avoidable from a common sense perspective, was unavoidable from the government bureaucrat's. Prior to Covid-19, the status quo in the United States was to not wear masks and there was no data to support a departure from this norm. Preserving the status quo was the natural instinct of those who sought to manage the public perception of the risk, rather than manage the risk itself. The initial argument for masks relied on the non-scientific influences of common sense and gut instinct, both anathema to government bureaucrats and academic researchers who fear being held responsible for a decision more than they care about getting the decision right.

The same bureaucratic logic and perverse incentive structure that led to the initial folly of discouraging mask use led to the later over-correction of focusing almost entirely on masks. Once "the science" revealed that masks were things, recommending—in fact, mandating—masks became safe haven for the government bureaucrat. For a period of time masks

---

[1]Discouraged because wearing a mask may scare others into thinking you are sick, or it may lull you into a false sense of security, or virus particles can get stuck in the mask.

weren't just a part of the solution, they became *the solution*, just months after they were initially written off as unnecessary and even harmful.

Almost a year later, in January 2021, Fauci applied NE Philly logic almost verbatim while advocating that people wear not just for one, but two masks. Fauci told the *New York Times*, "If you have a physical covering with one layer, you put another layer on, it just makes common sense that it likely would be more effective." Common sense indeed: a mask is a thing, two masks are two things, and two is better than one. Yet it took almost a year to come to this realization. By that time, it was too late. When the virus was still somewhat contained, one (or two) masks can possibly help slow the spread in the initial stages; they can't kill the virus once it's already spread throughout the population.

Fauci's conversion to common sense came at significant cost—just not to Fauci. By early February 2021, the United States had the highest number of Covid-19 deaths of any country in the world by a factor of two, it had also experienced a year of economic, political and societal turmoil, due in no small part to the collateral damage of prolonged lockdowns, economic uncertainty and overbearing mandates put in place to counter the fallout of the initially laissez faire response. Fauci, meanwhile, remained the face of U.S. government response, his influence over U.S. policy having increased substantially in the Biden administration.

3   Data-Driven Drones

[Second Axiom of Naive Probabilism] *Until proven otherwise, assume that the future will resemble the past.*

At first, there was no data that masks work, no data that travel restrictions work, no data of human-to-human transmission, and so on. Indeed, there was no data for anything. And in the absence of data, the naive decision protocol is to maintain the status quo, in strict adherence to Axiom 3.

There's some validity to Axiom 3—the future often resembles the past, and in most cases there's little or no harm in assuming that it will—but as with all things naive, blind adherence to this axiom can have ruinous consequences. We already saw the consequences of this assumption in the case of masks. But the logic was applied much more widely in advocating against other precautionary measures to prevent the early spread of Covid-19.

Flights from China to the U.S. were shut down in late January—much to the chagrin of the Naive Probabilist—but flights from Europe continued until mid-March, long after outbreaks had gotten out of control in Italy and elsewhere in Europe. American universities were among the earliest and biggest super-spreaders of Covid-19, with students and faculty regularly traveling to and visiting from all parts of the world. Despite their central status, both as vectors of disease spread and as leaders of the response, universities set the tone for delayed and naive action. Stanford waited (patiently) to close campus until a faculty member contracted the disease, and others followed suit, with most remaining open for more than a week after the Stanford shutdown, citing "no confirmed cases" on their own campuses. (Remember Axiom 3: the future resembles the past. There have been no past cases of Covid-19 on campus, and therefore one should assume there will be no future cases until proven otherwise.) University of Chicago was among the most lackadaisical, announcing on March 12 that it would wait until March 30 to begin remote activity, not wanting the virus to disrupt its final exam schedule. All the while, UChicago administrators maintained that there were no confirmed cases of anyone affiliated with the university, even though there were cases in its immediately surrounding neighborhood.

The delayed responses of the above and many other universities is a microcosm of the Naive Probabilist's worldview: whatever can't be explained in terms of something that happened in the past is speculative, non-scientific and unjustifiable. This argument was put forward by John Ioannidis in mid-March 2020, as the pandemic outbreak was already spiraling out of control. Ioannidis wrote that Covid-19 wasn't a "once-in-a-century pandemic", as many were saying, but rather a "a once-in-a-century data fiasco" [5]. Ioannidis's main argument was that we knew very little about the disease, its fatality rate, and the overall risks it poses to public health; and that in face of this uncertainty, we should seek data-driven policy decisions. Until the data was available, we should assume Covid-19 acts as a typical strain of the flu (a different disease entirely). Under Ioannidis's analysis, there were scenarios under which Covid-19 would be much more fatal than the annual flu and other scenarios under which it is much less fatal. The prudent approach, according to Ioannidis, was to delay response until we had a more definitive answer.

In academic circles, Ioannidis's article was regarded not as a misguided and potentially disastrous recommendation by an influential scientist but as "good contrarian writing", as epidemiologist Marc Lipsitch described it [7]. As a member of the academic establishment, Lipsitch was in the unfamiliar situation in which his advice had real consequences in real time. Rather than focus on the problem at hand, he instead focused on the academic exercise of "starting a discussion" about whether there was "sufficient data" to draw a conclusion about the dangers of Covid-19. For their participation in this thought exercise, Ioannidis and Lipsitch, like Fauci above, gained greater personal exposure and influence while delaying necessary action on the pandemic.

The problem with the "lack of evidence" argument was that there was, in fact, plenty of evidence well before the virus was spread throughout the United States [2]. China had locked down a city of 10 million; Italy had locked down its entire northern region, with the entire country soon to follow. There was overwhelming evidence, in fact worldwide consensus, that the virus was novel. The drastic measures taken in China, Italy and elsewhere were enough to conclude that (i) the virus was spreading fast and (ii) the medical communities in those countries had no idea how to treat it. That's data. It's not the kind of data that is curated by a lab or organized in a spreadsheet. But it's plenty of information to act on.

4   Sunstein–Costanza Fallacy

[Third Axiom of Naive Probabilism] *In the presence of uncertainty, derive wisdom from ignorance.*

Think of all the times you've been wrong in the past. If you could have just done the opposite, you'd have been right. This, in a nutshell, is the *Sunstein–Costanza Fallacy*, named after George Costanza, a fictional character from the 1990s sitcom *Seinfeld*, and Cass Sunstein, a real-life academic who repurposed Costanza's sitcom fallacy as a catchall strategy for dealing with uncertainty in real world problems. The Sunstein–Costanza fallacy takes the Naive Probabilist's credo—the future resembles the past (until proven otherwise)—one step further. It derives knowledge out of ignorance by analyzing past situations in which we've been ignorant, seeing how we reacted then, observing that those reactions were sub-optimal in hindsight, and concluding that we should do the opposite of what our instincts tell us in the present situation.

On February 29, 2020, there were 1,129 confirmed cases in Italy, up from 79 cases a week before. On that same day, Cass Sunstein chided Americans concerned over Covid-19 in a *Bloomberg* column:

> "At this stage, no one can specify the magnitude of the threat from the coronavirus. But one thing is clear: A lot of people are more scared than they have any reason to be."

The two sentences are contradictory on their own—if the magnitude of threat is unknown, then how does Sunstein know the level of fear is unreasonable?—but that's not the worst part of Sunstein's commentary. Rather than acknowledge that the uncertainty about the magnitude of the threat warrants a precautionary response to the pandemic—only after being definitively proven wrong did Sunstein later reverse course and advocate for precaution in another *Bloomberg* article three weeks later[2]—Sunstein applied Axiom 4 to derive wisdom from his ignorance about the magnitude of the threat. Sunstein diagnoses anyone concerned about Covid-19 with a cognitive defect known as 'probability neglect', which he defines as the tendency to fixate on very low probability, but highly impactful outcomes (good or bad) instead of focusing on what is most likely. With Covid-19, Sunstein argues that focusing on the possibility (small according to Sunstein) of a global pandemic, instead of the more likely outcome that the

---

[2] "This Time the Numbers Show We Can't Be Too Careful".[3]

disease is about as dangerous as the flu, is a prime example of probability neglect.

Applying Axioms 2-4, Sunstein argues first that (i) most previous times of mass panic (e.g., the H1N1 outbreak) fell far short of the extreme outcome that caused the panic and, therefore, (ii) future instances of panic are also likely to fall short of their projected worst-case scenario. In other words, since we've overreacted unnecessarily (according to Sunstein) in the past, we are most likely overreacting unnecessarily this time. Since our future panic resembles our past panic, we can apply Axiom 4 and derive wisdom from our ignorance about the specific situation at hand: Don't panic over Covid-19, not because of anything we know about it, but because of our extreme ignorance. When we've been ignorant in the past, we've over-reacted, therefore we're likely to be over-reacting again this time.

The above reasoning is an example of the *Sunstein–Costanza Fallacy*. In the episode of *Seinfeld* called "The Opposite", George Costanza reasons that he can improve his decision making by doing the opposite of what his instincts tell him.

> Costanza: "It became very clear to me sitting out there today, that every decision I've ever made, in my entire life, has been wrong. My life is the opposite of everything I want it to be. Every instinct I have, in every aspect of life, be it something to wear, something to eat ... It's all been wrong."
>
> Seinfeld: "If every instinct you have is wrong, then the opposite would have to be right."
>
> (*Seinfeld*, Episode #86, "The Opposite". https://www.seinfeldscripts.com/TheOpposite.htm)

Channeling Costanza, Sunstein's argument wasn't based on data or evidence, but rather an appeal to ignorance. A call to do the opposite of what we've done before. As Sunstein argues, we've been in this kind of situation before: before every big storm, the supermarket shelves empty; during the Zika virus, Ebola outbreak or swine flu, people cancel travel. In all of the previous cases, we can assess in hindsight that the virus either wasn't as deadly or contagious as originally feared. Sunstein concludes that that we were irrational to be concerned in those situations, and since we are now in a similar state of uncertainty relative to Covid-19 as we once were relative to Zika, swine flu, or a snowstorm, we're irrational to have such concerns over Covid-19. Indeed, because I survived one pull of the trigger in Russian roulette, I was irrational to think that I could have died on the first round, and thus also irrational to think that the next pull poses risk.

5   The Freeroll Effect and Covid-19

In gambling terms, a *freeroll* is a bet that can be won but not lost. In the worst case scenario, the bet breaks even. For the gambler, there is no risk, only upside. For the party on the other side, there's no upside, only risk. The gambler is said to be *freerolling*; the other side is *getting freerolled*.[4]

Outside of gambling, the *Freeroll Effect* arises when an influential party is allowed to reap the rewards of its influence without suffering the consequences. Businesses deemed "too big to fail" are bailed out rather than suffer the consequences of their neglect for excessive risks. In academia, "peer review" indemnifies scientists against publication of flawed findings on the grounds that the work has been vetted by expert peer reviewers in their field. In all cases, the ones responsible for the decisions are inoculated from the ill-effects of those decisions. And in the case of public policy making, such as pandemic response, those who influence the policy enjoy the benefits of their influence while facing none of the negative consequences when those policies backfire.

In the case of Covid-19, the above stories of Fauci, Ioannidis and Sunstein offer three high profile instances of the impact of Naive Probabilism and the Freeroll Effect. All three were influential in the early and ongoing public perception and response to the pandemic, and all three were wrong in their initial recommendations over impactful decisions. Despite their negative impact, Fauci, Ioannidis and Sunstein enjoyed increased influence as the pandemic wore on: Fauci's influence and public appearances increased after the transition to the Biden administration; Ioannidis gained increasing media attention throughout March and April 2020; and Sunstein was named to chair a Covid-19 technical advisory group at the World Health Organization. All were benefactors of the Freeroll Effect: they were indemnified (by society) against the large-scale risks of their mistakes, while they enjoyed the benefits of public influence.

In any complex system with competing incentives, there is usually someone in the position of a Fauci, an Ioannidis or a Sunstein, who standard to benefit at the expense of the rest. The specific individuals mentioned above are by no means unique in their benefiting from the Freeroll Effect. Except for their serving as vehicles of potentially widespread harm, they need not be particularly villainous or mean-spirited. The Freeroll Effect arises almost any time there is an asymmetric sharing of risk and of consequences between those who influence consequential decisions and those who are most impacted by those decisions.

---

[4]Note that *freeroll* is not synonymous with *arbitrage*. Arbitrage is a financial concept describing opportunities for risk-free profits in financial markets. Most simply, arbitrage exists when it is possible to simultaneously buy and sell an asset at prices that generate profit. The arbitrageur's profits need not be at the expense of the buying or selling counterparties. The buyer and seller, as participants in a market, voluntarily offered to buy and sell at specific bids and asks, and therefore willingly entered into the transaction. The party being freerolled rarely enters voluntarily into such an arrangement.

6 The Naive Probabilist

The Naive Probabilist believes that all decisions under uncertainty boil down to probability calculations; that sound decision making is a just math problem, a simple matter of collecting data and "turning the crank" given to us by probability theory. As any "good Bayesian", the Naive Probabilist updates based on new information, but often waits too long to act on that information, or fails to recognize information that doesn't come in the form of a well-manicured dataset.

The Naive Probabilist believes that the future is like the past, the house always wins, that all available information is "priced in", debts always get paid, that the real world obeys the theory (on average), that deviations from theory indicate a problem with the real world, that good intentions are more important than good results, that ignorance begets knowledge, and above all, that all models are wrong, but some are useful. In practical terms, he (or she) waits until the car is buried in the ditch to put on his seatbelt. At the poker table, he waits until the cards are tabled to fold the worst hand. He waits until the disease is a six-continent pandemic, the global economy is in disarray, and the hospitals are full to determine, with certainty, that Covid-19 is a public health threat.

To be clear, Axioms 2-4 are axioms to the Naive Probabilist, but fallacies to everyone else. Complex problems call for simple, actionable solutions (i.e., a mask is a thing); the past doesn't repeat indefinitely (i.e., Covid-19 was never the flu); and ignorance is not a form of wisdom (i.e., contrary to nudge theory and behavioral economics, de-training our instincts isn't a sound approach to decision making). The Naive Probabilist's primary objective— to be accurate with high probability rather than to protect against high consequence, but low probability outcomes—goes against common sense principles of decision making in severe uncertainty, severe consequence situations. As I and others have written elsewhere, in the presence of severe uncertainty, precautionary principles, common sense and basic survival instincts should predominate [1,4,8,9]. In such situations, accuracy is the least of our concerns.

With that said, I stress that the hallmark of Naive Probabilism is naiveté, not ignorance, stupidity, crudeness or other such base qualities. In fact, the typical Naive Probabilist lacks not knowledge nor refinement, but the experience and good judgment that comes from making real decisions with real consequences in the real world; see Sections 2-4 for three examples. Far from ignorant, the most prominent naive probabilists are recognized (academic) experts in mathematical probability, or relatedly statistics, physics, psychology, economics, epistemology, medicine or so-called decision sciences. Beyond their sterling credentials, the best known (and most dangerous) naive probabilists are quite sophisticated, skilled in the art of influencing public policy decisions without suffering from the risks those policies impose on the rest of society.

Thanks to the Freeroll Effect, naive probabilists continue to influence important decisions with far-reaching impact on the way society operates, government runs, and the economy progresses. The above commentary on Naive Probabilism and the Freeroll Effect in Covid-19 is a cautionary tale of the widespread influence of Naive Probabilism throughout society, science, academia, business, medicine, finance, law and politics. Naive Probabilism is responsible for prospect theory, the GRE, "libertarian paternalism", Mitt Romney, ill-fated attempts to "solve" the replication crisis, and the early and ongoing response to the Covid-19 pandemic in the U.S. and throughout the West.

Naive Probabilism persists with the help of the Freeroll Effect, and also because in many, non-complex domains the tenets of Naive Probabilism have little or no major impact. But as we've seen in this case study of Covid-19, Naive Probabilism is most noticeable and most detrimental in complex systems, where its core axioms are the most wrong and have their most severe consequences. Drawing from this case study, we may better recognize the occurrence of Naive Probabilism and the Freeroll Effect in other domains, and hopefully mitigate or entirely avoid similar catastrophes in the future, whether economic, political, or health-related.

7   The Enlightened Probabilist

In contrast to the Naive Probabilist stands the Enlightened Probabilist, who unlike Sunstein and Co. understands that decision making is situational. Context is everything. The Enlightened Probabilist knows the theory inside-out, but isn't blinded by it. He realizes that the theory applies only under specific circumstances. Decision making under uncertainty is practical, emotional and psychological.

The Enlightened Probabilist adheres to common sense. He (or she) wears a seatbelt, looks both ways before crossing (even when the light is green), locks his (or her) doors, keeps cash on hand ("dry powder"), stores extra ammo (dry powder), and avoids dark alleys. There are times when the Englightened Probabilist neglects to do these things, and no bad comes of it. He forgets to wear a seatbelt—no accident; forgets to look before crossing—no car coming; forgets to lock the door—no robbery; runs out of cash—didn't need it; runs out of bullets—no altercation; walks down a dark alley—nobody there. The Enlightened Probabilist knows ahead of time that these precautions safeguard against things that are all unlikely to happen, but also knows that probability isn't just about what's "likely".

The Naive Probabilist denies that what's "rational" for one person may be irrational for another, and that the right decision in one context may be the wrong decision in another. The Naive Probabilist rejects the Enlightened Probabilist's Mantra:

The Enlightened Probabilist's Mantra

When gambling, think probability.
When hedging, think plausibility.
When preparing, think possibility.
When this fails, stop thinking. Just survive.

To the extent that the Naive Probabilist follows the Mantra, he gets stuck on the first line: everything comes down to a gamble, an expected value, or a utility calculation. The Enlightened Probabilist knows that the Mantra applies from bottom up. First, survive (avoid ruin). Second, maintain (avoid loss). Third, thrive (win), time and resources permitting.

The Enlightened Probabilist would much rather be alive than look smart. To quote Warren Buffet, in the context of investing, "In order to succeed, you must first survive." To invest successfully requires capital; and to have capital one mustn't be broke. But the principle applies much more widely, at individual and societal levels. To do anything, one must first survive, and survival isn't a matter of probability, but of possibility. Seatbelts, locks, "dry powder", dry powder, extra food and water all guard against the *possibility* of a crash, break-in, economic hardship, altercation, famine or drought, no matter how unlikely any of them may be.

Beyond survival, the Enlightened Probabilist hates being squeezed, stays liquid, buys insurance, hedges his bets. He doesn't assume that *his* best explanation is *the* best explanation, or that his understanding incorporates all available information. He realizes that he is error prone, and therefore needs to protect against not only the most likely scenarios but also any additional *plausible* scenarios, especially those that would lead to substantial harm. At this stage, the Enlightened Probabilist isn't trying to win the most, but to lose the least.

Only after shoring up survival and protecting against excessive loss does the Enlightened Probabilist even consider profiting, winning, or "being right" in any sense. At this point the Enlightened Probabilist has the luxury of considering the *probability* of the outcomes, but it's a long road to get there.

The many levels of risk and uncertainty

- (easy) *Theory*: what they taught in school.

- (hard) *Practice*: what you learned in the schoolyard.

- (harder) *Psychology*: how much you can handle.

- (hardest) *Ethics*: who you really are.

An understanding of the many levels of risk and uncertainty distinguishes Naive from Enlightened. For the Naive Probabilist, the theory is the only part. For the Enlightened Probabilist, the theory is just the easy part. Indeed, there are situations in which probability is the right concept and probability theory is the correct framework for decision making: when the probabilities are known or reasonably well estimated, and when the payoffs (especially downside) is bounded. For all the reasons discussed previously, the early days of Covid-19 was not such a situation, and the major individuals and organizations who influenced pandemic response were far from enlightened, and far from ethical.

A crucial step in bringing the theory to practice is to identify all the ways that the theory fails to apply. Hypotheses of uniformity, independence, infinite-population, and large-scale asymptotics all assume behaviors that don't exist in the real world. Beyond the practical, there are the emotional and ethical aspects of decision making. Decisions under uncertainty impose a psychological burden: the right decision may go badly wrong, and the Enlightened maintains composure under this situation. They also impose an ethical imperative: one mustn't impose risks upon others which they don't subject themselves. Refer to Section 5 for discussion of how this ethical imperative was violated in the Covid-19 response, and is regularly violated in large-scale decisions under uncertainty in complex systems.

Naivete repeats itself

The above case studies have the benefit of hindsight to illustrate the impact of Naive Probabilism on early Covid-19 response. As I write this a year after the initial outbreak of March 2020, many of the *post hoc* over-reactions (e.g., lockdowns, mandates, travel restrictions) remain in force in some parts of the United States. Meanwhile a number of new interventions, vaccines chief among them, are being implemented at large scale. Several Covid-19 vaccines have begun distribution on an Emergency Use Authorization (EUA) by the U.S. Food and Drug Administration (FDA). For some people, the vaccine will be a lifesaver. But for others, it presents another source of severe uncertainty and unwanted risk.

Much like early detractors of precautionary measures, the Naive response to concerns over vaccine risk has been to dismiss context dependence in decision making—what's rational for one person may be irrational for another—in favor of a public outreach initiative that seeks 100% adoption. June Raine, CEO of the United Kingdom's Medicines and Healthcare products Regulatory Agency (MHRA) assured the public that "the benefits outweigh any risk" of Covid-19 mRNA vaccines. Indeed, for some people, the benefits do outweigh any risk, as Raine suggests. For many others, however, the untold risks far outweigh any benefits. Naive Probabilism arises from the belief that risks are uniformly shared across everyone, and that there is a unique correct decision to every challenge. In doing so, Naive Probabilists replace one relatively known risk (Covid-19) with another much lesser known one (uncertainty of the vaccine risks).

As with the initial handling of masks and precautions, efforts to squash vaccine concerns have only shed more doubts about the credibility of leading scientists. In the minds of those concerned, the Covid-19 vaccines approved for EUA in the U.S. were developed at "warp speed"[5] to treat a novel virus using a technology (messenger RNA, mRNA) which has never before been approved for use in humans. Compounding this natural skepticism is the fact that many of the same people who denied the early risks of Covid-19 are now dismissing potential risks of widespread vaccination, and doing so in an admittedly deceitful way. When polls suggested that about half of Americans would refuse to get the vaccine due to concerns over its safety, Drs. Fauci and Adams made a number of media appearances to assure the public that the vaccines were safe. Fauci told the *New York Times*:

> "When polls said only about half of all Americans would take a vaccine, I was saying herd immunity would take 70 to 75 percent," Dr. Fauci said. "Then, when newer surveys said 60 percent or more would take it, I thought, 'I can nudge this up a bit,' so I went to 80, 85." [6]

The above observations about masks, data, rationality, and risk assessment of vaccines highlight the failed thought process underlying all of these decisions. The problem with these decisions isn't that they were "right" or "wrong" with the benefit of hindsight—whether current vaccine recommendations prove beneficial remains unknown—but that they originated from a naive understanding of probability and its proper place in decision making under severe uncertainty.

Bibliography

[1] The Precautionary Principle. https://www.fooledbyrandomness.com/ PrecautionaryPrinciple.html.

[2] H. Crane. A fiasco in the making: More data is not the answer to the coronavirus pandemic. *Researchers.One*.

[3] H. Crane. The Fundamental Principle of Probability. *Researchers.One*, 2018 (https://www.researchers.one/article/2018-08-16).

[4] H. Crane. Naive Probabilism. *Researchers.One*, 2020 (https://researchers.one/articles/20.03.00003).

[5] J. Ioannidis. A fiasco in the making? As the coronavirus pandemic takes hold, we are making decisions without reliable data. *STAT*, March 17, 2020 (https://www.statnews.com/2020/03/17/a-fiasco-in-the-making-as-the-coronavirus-pandemic-takes-hold-we-are-making-decisions-without-reliable-data/).

---

[5]Supported by the U.S. Department of Defense "Operation Warp Speed".
[6]https://www.nytimes.com/2020/12/24/health/herd-immunity-covid-coronavirus.html

[6] Howard, et al. The effectiveness of face masks to prevent SARS CoV-2 trans-
    mission: A summary of the peer-review science. *Proceedings of the U.S. National
    Academy of Sciences*, 2020.

[7] M. Lipsitch.   We know enough now to act decisively against
    Covid-19. Social distancing is a good place to start.   *STAT*, March
    18, 2020 (https://www.statnews.com/2020/03/18/we-know-enough-now-to-act-
    decisively-against-covid-19/).

[8] J. Norman. Global Decentralization for Risk Mitigation and Security. *Re-
    searchers.One*, 2020 (https://www.researchers.one/article/2020-02-15).

[9] J. Norman, Y. Bar-Yam, and N. Taleb. Systemic Risk of Pandemic via Novel
    Pathogens – Coronavirus: A Note, 2020 (https://necsi.edu/systemic-risk-of-
    pandemic-via-novel-pathogens-coronavirus-a-note).

On Naive Forecasts for Fat-Tailed Variables

Nassim Nicholas Taleb[1,2,*], Yaneer Bar-Yam[3] and Pasquale Cirillo[4,5]

[1]Universa Investments
[2]Tandon School of Engineering, New York University
[3]New England Complex Systems Institute
[4]Institute for the Future, University of Nicosia
[5]S-T-A-T-S GmbH, Switzerland.
[*]Corresponding author: nnt1@nyu.edu

We discuss common errors and fallacies when using naive "evidence based" empiricism and point forecasts for fat-tailed variables, as well as the insufficiency of using naive first-order scientific methods for tail risk management. We use the COVID-19 pandemic as the background for the discussion and as an example of a phenomenon characterized by a multiplicative nature, and what mitigating policies must result from the statistical properties and associated risks. In doing so, we also respond to the points raised by Ioannidis et al.(2020)

July 27, 2020. We thank Pierre Pinson and Spyros Makridakis for helping organize this discussion, and John Ioannidis for his gracious engagement.

Main Statements

(i) Forecasting single variables in fat-tailed domains is in violation of both common sense and probability theory.

(ii) Pandemics are extremely fat-tailed events, with potentially destructive tail risk. Any model ignoring this is necessarily flawed.

(iii) Science is not about making single points predictions but about understanding properties (which can *sometimes* be tested by single point estimates and predictions).

(iv) Sound risk management is concerned with extremes, tails and their full properties, and not with averages, the bulk of a distribution or naive estimates.

(v) Naive fortune-cookie evidentiary methods fail to work under both risk management and fat tails, because the absence of raw evidence can play a large role in the properties.

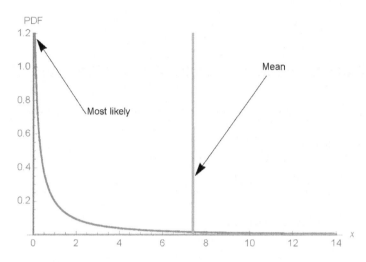

Figure 1: A high variance log-normal distributions. 85% of observations fall below the mean. Half the observations fall below 13% of the mean. The log-normal has milder tails than the Pareto which has been shown to represent pandemics.

(vi) There are feedback mechanisms between forecast and reaction that affects the validity of some predictions.

(vii) Individuals risks fail to translate into systemic risks under multiplicative processes.

(viii) One should never treat the "costs" of mitigation without taking into account the costs of the disease, and in some cases naive cost-benefit analyses fail (for sure when statistical averages are nonconvergent or invalid for tail risk purposes).

(ix) Historically, in the aftermath of the Great Plague, economies were less fragile to pandemics, equipped to factor-in effective mechanisms of containment (quarantines) in their operating costs. It is more cogent to blame overoptimization than reaction to disease.

The article is organized at three levels. First, we make general comments around the nine points in the Main Statements, explaining how single point forecasts is an unscientific simplification incompatible with processes with richer properties. Next we go deeper into the technical arguments. Finally we address specific points in Ioannidis et al. [16] and answer their arguments concerning our piece.

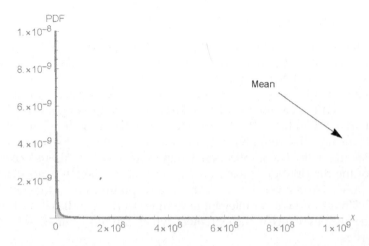

Figure 2: A Pareto distribution with a tail similar to that of the pandemics. Makes no sense to forecast a single point. The "mean" is so far away you *almost* never observe it. You need to forecast things other than the mean. And most of the density is where there is noise.

Commentary

Both forecasters and their critics are wrong

At the onset of the COVID-19 pandemic, many research groups and agencies produced single point "forecasts" for the pandemic—most relied upon trivial logistic regressions, or upon the compartmental SIR model, sometimes supplemented with cellular automata, or with agent-based models assuming various social rules and behaviors. Apparently, the prevailing idea is that *producing a single numerical estimate* is how science is done, and how science-informed decision-making ought to be done: bean counters producing precise numbers. And always within a narrowly considered set of options identified by the researchers.

Well, no. That is not how "science is done", at least in this domain, and that is not how informed decision-making should develop.

Furthermore, subsequently and ironically, many criticized the plethora of predictions produced, because these did not play out (no surprise there). This is also wrong, because both forecasters (who missed) and their critics (complaining) were wrong. Indeed, forecasters would have been wrong anyway, even if they had got their predictions right. In fact, as we will clarify throughout this article, 1) in some domains (i.e. under fat tails) naive forecasts are poor descriptors of a system (hence highly unscientific), even when they might appear reasonable; 2) for some functions (risk management related), or some classes of exposures (systemic ones), these forecasts are extremely misplaced.

Statistical attributes of pandemics

Using tools from extreme value theory (EVT), Cirillo and Taleb [6] have recently shown that pandemic deaths are *patently* fat-tailed[1]–a fact some people like Benoit Mandelbrot (or one of the authors, in *The Black Swan* [32]) had already guessed, but never formally investigated. Even more, the estimated tail parameter α is smaller than 1, suggesting an apparently infinite risk [6], in line with destructive events like wars [4,5], and the so-called "dismal" theorem [39]. Pandemics do therefore represent a source of existential risk. The implication is that much of what takes place in the bulk of the distribution is just noise, according to "the tail wags the dog" effect [6,33]. And one should never forecast, pontificate, or theorise from noise! Under fat tails, all relevant and vital information lies in fact in the tails themselves (hence in the extremes), which can show remarkably stable properties.

Remark 1 (Observed events vs Observed Properties) *Random variables with unstable (and uninformative) sample moments may still have extremely stable and informative tail properties, centrally useful for robust inference and risk taking. Furthermore, these reveal evidence.*

This is the central problem with the misunderstanding of **The Black Swan** *[32]: some events may have stable and well-known properties, yet they do not lend themselves to prediction.*

Fortune-cookie evidentiary methods

In the early stages of the COVID-19 pandemic, scholars like Ioannidis [14] suggested that one should wait for "more evidence" before acting with respect to that pandemic, claiming that "we are making decisions without reliable data".

Firstly, there seems to be some probabilistic confusion, leading towards the so-called delay fallacy [13]: "if we wait we will know more about X, hence no decision about X should be made now."

In front of potentially fat-tailed random variables, more evidence is not necessarily needed. Extra (usually imprecise) observations, especially when coming from the bulk of the distribution, will not guarantee extra knowledge. Extremes are rare by definition, and when they manifest themselves it is often too late to intervene. Sufficient –and solid – evidence, in particular for risk management purposes, is already available *in the tail properties themselves.* An existential risk needs to be killed in the egg, when it is still cheap to do so. Events of the last few months have shown that

---

[1]A non-negative continuous random variable X has a fat-tailed distribution, if its survival function $S(x) = P(X \geq x)$ is regularly varying, formally $S(x) = L(x)x^{-\alpha}$, where $L(x)$ is a slowly varying function, for which $\lim_{x \to \infty} \frac{L(tx)}{L(x)} = 1$ for $t > 0$ [7,9,10]. The parameter α is known as the tail parameter, and it governs the fatness of the tail (the smaller α the fatter the tail) and the existence of moments ($E[X^p] < \infty$ if and only if $\alpha > p$).

waiting for better data has generated substantial delays, causing thousands of deaths *and* serious economic consequences.

Secondly, unreliable data[2]–or any source of serious uncertainty–should, under some conditions, make us follow the "paranoid" route. More uncertainty in a system makes precautionary decisions more obvious. If you are uncertain about the skills of the pilot, you get off the plane *when it is still possible to do so*. If there is an asteroid headed for earth, should we wait for it to arrive to see what the impact will be? We might counter that there were asteroids in the past that had devastating impacts, and besides we can calculate the physics. The logical fallacy runs deeper: "We did not see this particular asteroid yet" misses the very nature of the power of science to generalize (and classify), and the power of actions to possibly change the outcome of events. Similarly, if we had a hurricane headed for Florida, a statement like "We have not seen this hurricane yet, perhaps it will not be like the other hurricanes!" misses the essential role of risk management: to take preventive actions, not to complain ex post. And if people take action boarding up windows, and evacuating, the claim "look it was not so devastating", that someone might afterwards make, should be considered closer to a lunatic conspiracy fringe than scientific discourse.

By definition, evidence follows and never precedes rare impactful events. Waiting for the accident before putting the seat belt on, or evidence of fire before buying insurance would make the perpetrator exit the gene pool. Ancestral wisdom has numerous versions such as *Cineri nunc medicina datur* (one does not give remedies to the dead), or the famous saying by Seneca *Serum est cavendi tempus in mediis malis* (you don't wait for peril to run its course to start defending yourself).

However, just as there are frivolous lawsuits there are frivolous risk claims and, as we will see further down, we limit these precautionary considerations to a precise class of fat tailed multiplicative processes –when there is systemic risk.

Remark 2 (Fundamental Risk Asymmetry) *For matters of survival, particularly when systemic, and in the presence of multiplicative processes (like a pandemic), we require "evidence of no harm" rather than "evidence of harm."*

Technical Comments

The Law of Large Numbers (LLN) and Evidence

In order to leave the domain of ancient divination (or modern anecdote) and enter proper empirical science, forecasting must abide by both evidentiary and probabilistic rigor. Any forecasting activity about the mean (or

---

[2]Many of those complaining about the quality of data and asking for more evidence before taking action, even in extremely risky situations, rarely treat the inputs of their predictive models as imprecise [4, 38], stressing them, and performing serious robustness checks of their claims.

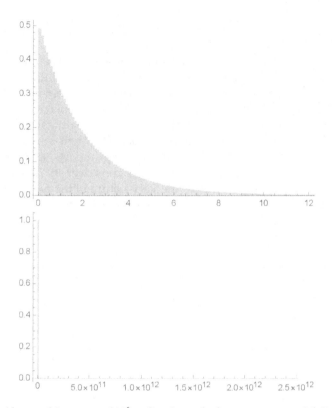

Figure 3: Above, a histogram of 10^6 realizations of r, from an exponential distribution with sole parameter $\lambda = \frac{1}{2}$. Below, that of $X = e^r$. We can see the difference between the two distributions. The sample kurtosis are 9 and 10^6 respectively (in fact it is infinite for the second) –all values for the second one are dominated by a single large deviations.

the parameter) of a phenomenon requires the working of the law of large numbers (LLN), guaranteeing the convergence of the sample mean at a known rate, when the number n of observations increases. This is surely well-known and established, except that some are not aware that, even if the theory remains the same, the actual story changes under fat tails.

Even in front of the most well-behaved and non-erratic random phenomenon, if one claimed fitness or non-fitness of a forecasting ability on the basis of a single observation ($n = 1$), he or she would be rightly accused of unscientific claim. Unfortunately, with fat-tailed variables that "$n = 1$" error can be made with $n = 10^6$. In the case of events like pandemics, even larger $n \to \infty$ can still be anecdotal.

Remark 3 (LLN and speed of convergence) *Fat-tailed random variables with tail exponent $\alpha \leq 1$ are simply not forecastable. They do not obey the LLN, as their* *theoretical mean is not defined, so there is nothing the sample mean can converge to. But we can still understand several useful tail properties.*

And even for random variables with $1 < \alpha \leq 2$, the LLN can be extremely slow, requiring an often unavailable number of observations to produce somehow reliable forecasts.

As a matter of fact, owing to pre-asymptotic properties, a conservative heuristic is to consider variables with $\alpha \leq 2.5$ as not forecastable in practice. Their sample mean will be too unstable and will require way too much data for forecasts to be reliable in a reasonable amount of time. Notice in fact that 10^{14} observations are needed for the sample mean of a Pareto "80/20", with $\alpha \approx 1.13$, to emulate the gains in reliability of the sample average of a 30-data-points sample from a Normal distribution [33].

Assuming significance and reliability with a low n is an insult to everything we have learned since Bernoulli, or perhaps even Cardano.

Also notice that discussing the optimality of any alarm system [3,18,31] trying to perform predictions on averages would prove meaningless under extremely fat-tails, i.e. when $\alpha \leq 2$, that is when the LLN works very slowly or does not work. In fact, even when the expected value is well-defined (i.e. $1 < \alpha < 2$), the non-existence of the variance would affect all the relevant quantities for the verification of optimality [8], from the size of the alarm to the number of correct and false alarms, from the probability of detection of catastrophes to the chance of undetected events. For all these quantities, the naive sample estimates commonly used would prove misleading. A solution could be the implementation on EVT-based approaches, possibly with the additional tools of [5] or [21], but at this stage nothing similar exists, to the best of our knowledge.

For this and other reasons specified later, the application of a non-naive precautionary principle [23] appears to be the viable solution in front of potentially existential risks.

Science is about understanding properties, not forecasting single outcomes

Figures 1 and 3 show the extent of the problem of forecasting the average (and so other quantities) under fat tails. Most of the information is away from the center of the distribution. The most likely observations are far from the true mean of the phenomenon and very large samples are needed for reliable estimation. In the lognormal case of Figure 1, 85% of all observations fall below the mean; half the observations even fall below 13% of the mean. In the Paretian situation of Figure 3, mimicking the distribution of pandemic deaths, the situation gets even worse: the mean is so far away that we will almost never observe it. It is therefore preferable to look at other quantities, like for example the tail exponent.

In some situations of fast-acting LLN, as (sometimes) in physics, properties can be revealed by single predictive experiments. But it is a fallacy to assume that a single predictive experiment can actually validate any theory; it is rather a single tail event that can falsify a theory.

Sometimes, as recently shown on the *International Journal of Forecasting* by one of the authors [34], a forecaster may find a single quantity that is actually forecastable, say the survival function. For n observations a tail survival function has an error of $o(\frac{1}{n})$, even when tail moments are not tractable, which is why many predict binary outcomes–as with the "superforecasting" masquerade. In [33], it is shown how–paradoxically–the more intractable the higher moments of the variable, the more tractable the survival function becomes. Metrics such as the Brier score are well adapted to binary survival functions, though not to the corresponding random variables. That is why survival functions are essentially useless for risk management purposes. In insurance, for instance, one never uses survival functions for hedging, but rather expected shortfalls–binary functions are reserved to (illegal) gambling.[3]

We do not observe properties of empirical distributions

A commentator (Andrew Gelman) [12] wrote "The sad truth, I'm afraid, is that Taleb is right: point forecasts are close to useless, and distributional forecasts are really hard."

The problem is actually worse. In fact, distributional forecasts are more than hard–and often uninformative. Building so-called empirical distributions by survival functions does not reveal tail properties since it will necessarily be censored and miss tail observations –those that under very fat tails (say $\alpha \leq 2$) harbor not most, but literally all of the properties [33]. In other words, probabilities are thin-tailed (since they are bounded by 0 and 1) but the corresponding payoff is not, so small errors in probability translate into large changes in payoffs. However, as further discussed in [6], the tail parameters are themselves thin-tailed distributed, hence reveal their properties rather rapidly. Simply, tail parameters extrapolate–while survival functions don't–and methods to measure the tail are quite potent.[4]

Uncertainty goes one way; errors in growth rates induce biases and massive fat tails for the quantity of interest

Consider the simple model

$$X_t = X_0 e^{r(t-t_0)},$$

---

[3]The main problem is that the conditional expectation is not convergent: $\lim_{K \to \infty} \frac{1}{K} E(X|X > K) > 1$, see [34] for a lengthy discussion.

[4]This also relates to the superforecasters masquerade mentioned earlier: building survival functions for tail assessments via sports-like "tournaments" as in [37], instead of using more rigorous approaches like EVT, is simply wrong and violates elementary probability theory.

where X_t represents the quantity of interest (say the number of fatalities in pandemics) between periods t_0 and t,

$$r = \frac{1}{(t - t_0)} \int_{t_0}^{t} r_s ds$$

and r_s is an instantaneous rate.

Using the histograms of r and X, Figure 3 shows something fundamental: a well-behaved distribution, that of r, may lead to an intractable one, that of X; furthermore, the more volatile r, the more downward-biased your observation of the mean of X.

Implication: one cannot naively translate between the rate of growth r and X_T, because errors in r could be small (but surely not zero), but their impact will be explosive on X, because of exponentiation.

Simply, if r is exponentially distributed (or part of that family), X will be power law. The tail α is a direct function of the variance: the higher the variance of r, the thicker the tail of X.

Remark 4 (Errors in Exponential Growth) *1) Errors in growth rates of a disease increase the fatness of tails in the distribution of fatalities.*
2) Errors in growth rates translate, on balance, into higher expected casualties.

We note that in the context of dynamical systems an exponential dynamics is defined as chaotic [19]. While the study of chaos often considers systems with fixed parameters and variable initial conditions, the same sensitivities arise due to variations in parameters; in this case, the value of contagion rate (R) and the social behaviors that affect it. Indeed this means that by changing human behavior, the dynamics can be strongly affected, thus allowing for the opening of opportunities for extinction.[5]

Never cross a river that is 4 feet deep on average

Risk management (or policy making) should focus on tail properties and not on the body of probability distributions. For instance, The Netherlands have a policy of building and calibrating their dams and dykes not on the average height of the sea level, but on the extremes, and not only on the historical ones, but also on those one can expect by modelling the tail using EVT, via semi-parametric approaches [7,9].

---

[5]One of the authors has shown [25] that with increasing global transportation there is a phase transition to global extinction with probability 1. This indicates that historical distributions don't account for the severity or frequency of current or future extreme events because the fat tailed distributions themselves are coalescing to unit probability extreme events over shorter time intervals due to global changes in societal behaviors. During this process the probability distributions for events in any time interval becomes progressively more weighted to large scale events. Thus historical decadal or century intervals between pandemics are inadequate descriptors of current risk.

Science is not about safety

Science is a procedure to update knowledge; and it can be wrong provided it produces interesting discussions that lead to more discoveries. But real life is not an experiment. If we used a p-value of .01 or other methods of statistical comfort for airplane safety, few pilots and flight attendants would still be alive. For matters that have systemic effects and/or entail survival, the asymmetry is even more pronounced.

Forecasts can result in adjustments that make forecasts less accurate

It is obvious that if forecasts lead to adjustments, and responses that affect the studied phenomenon, then one can no longer judge these forecasts on their subsequent accuracy. Yet the point does not seem to be part of the standard discourse on COVID-19.

By various mechanisms, including what is known as Goodhart's law [30], a forecast can become a target that is gamed by participants–see also the Lucas' critique applying the point more generally to dynamical systems. In that sense a forecast can be a warning of the style "if you do not act, these are the costs" [6].

More generally, any game theoretical framework has an interplay of information and expectation that causes forecasts to become self-canceling. The entire apparatus of efficient markets–and modern economics–is based on such self-canceling aspect of prediction, under both rational expectations and an arbitrage-free world.

Remarks Specific to Ioannidis et al.

Systemic risks vs individual risks

A fundamental problem, in both [15] and [16], lies in ignoring scaling: systemic risks do not resemble (even qualitatively) individual risks. The macro and the micro-properties of contagious events, given their infective multiplicative nature, don't map directly onto one another.

Ioannidis et al. [15] write: "the average daily risk of dying from coronavirus for a person <65 years old is equivalent to the risk of dying driving a distance of 13 to 101 miles by car per day during that COVID-19 fatality season in 17 of the 24 hotbeds (...) For many hotbeds, the risk of death is in the same level roughly as dying from a car accident during daily commute."

Even if Ioannidis et al.'s computation were to hold true for one individual (it does not), conditionally on an excess of 10^3 of such individuals dying, the probability that the cause of death is COVID-19 and not a car accident

---

[6]For instance Dr. Fauci's warning that the number of (verified) infections could reach 100K per day (*New York Times*, June 30, 2020) should not be interpreted as a forecast to be judged according to its accuracy; rather a signal about what could happen should one avoid taking action.

converges to 1. When you die of a contagious disease, people around you are at risk of contagion, and they can then infect other people, in a cascading effect. It is quite elementary: car accidents are not contagious, while COVID-19 is. You cannot conflate the two objects: one is additive in the aggregate, the other is multiplicative. In [33], it has been shown that this is a severe error, leading to macroscopic blunders[7].

Remark 5 (Scaling of Probabilities) *Under multiplicative effects the risks for a collective do not scale up from the risks of an individual. Trivially, systemic risks can be extreme, where the individual ones are low, or vice-versa.*

Trade-offs and Ergodicity

One could say: panic saves lives, but at what economic price? Let us put aside ethical arguments, and answer it, ignoring for a moment the value of human life.

The fact is that some classes of (systemic) risks require being killed in the egg, also from an economic point of view. The good news is that there are not so many–but pandemics as we said fall squarely within the category.

The "dismal" theorem [39] mentioned earlier tells us that it is an error to use trade-off analysis under existential risk. There have been many proofs of similar arguments on grounds of ergodicity, well-known by insurance companies since Cramèr: simply, you cannot use naive B-school costs-benefit analyses for Russian roulette, because of the presence of an absorption barrier [32]. But one should not blame Ioannidis et al. [15] for this error in reasoning: it has been shown to be unfortunately prevalent in the decision-science literature [24].

Remark 6 (Ruin Problems) *Traditional cost-benefit analysis fails to apply to situations where statistical averages are unreliable, if not invalid.*

Moreover, it is not correct to assume, more or less implicitly, that a disease brings no or little costs, while mitigation is burdensome. There are indeed severe nonlinearities at play.

First of all, risk is beyond the simple and direct disease-specific mortality rate. In fact, letting the disease run above a certain threshold would compound its effect (in an explosive manner), because of the saturation of services, causing for example the displacement of other patients, many in potentially critical conditions; something that we have seen happening in the Region of Lombardy in Italy, in New York City, and elsewhere for several weeks during the spring of 2020 [29]. Furthermore, for survivors the illness itself represents a large economic drain, be it only from lost working hours, not counting the costs of hospitalization. And for every severe

---

[7]Note that this is also a typical example of "size fallacy," in which different risky events are compared just on the basis of their probabilities of occurrence, without caring about their different nature [13].

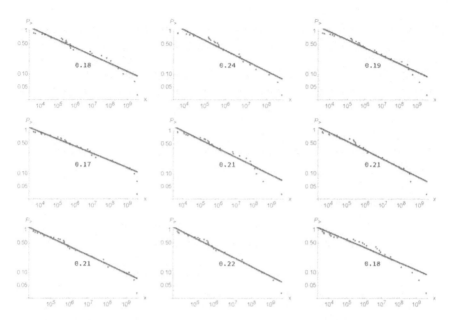

Figure 4: Zipf plots (log-log plots of the empirical survival function $P_>$) for nine random selections of 30 out of the 72 pandemics in [6]. The number in the center represents the a naive (OLS) estimate of the tail parameter α, readable as the absolute slope of the red negative line. The values of α appear to be stable notwithstanding the sampling, signalling the robustness of the approach and the inconsistency of the "selection bias" critique. The values are also in line with the more rigorous EVT-based findings of [6].

infection, there is an unspecified number of morbidities, with unknown (but definitely larger than zero) additional mortality and long term costs for the health system [1, 29], as it has been the case for other diseases like SARS [22].[8]

Remark 7 (False Dichotomies) *One should not treat the economy and the disease as separate independent items, particularly by viewing a naive trade-off between economic costs and pandemic mitigation.*

Moreover, never underestimate consumers' (nonlinear) behavior. When risks are visible (and a pandemic definitely is), people tend to modify their

---

[8]**Geronticide:** This discussion does not even cover the ethical discussion of trade-offs and their inapplicability in some domains, perhaps the most central discussion. At what price will you kill your parents/grandparents? A million dollars? Ten million? A billion? Furthermore, the fact that older people are more vulnerable to the disease brings considerations of geronticide (senicide): one misses that the silver rule [32] commands treating older generations under a moral liability, as one wishes to be treated by the next generation. Letting the disease run through older generations violates the interdicts on geronticide and inter-generational obligations. The fact that your parents did not sacrifice their own parents creates an obligation to not sacrifice them; your children will spare you in turn, under the same rule.

behavior, rationally or not, also switching to alternatives, with nonlinear effects on the businesses concerned [26]. This is the reason why the airline industry in the U.S. manages to have fewer than 1 fatal crash in 25×10^6 flights (and aims at an even more favorable ratio). One may claim that it is irrational to spend so much of our resources mitigating plane crashes, but airline companies know that, in case of fewer checks and efforts, consumers would then probably switch to other companies, if not directly to other types of transportation.

Take the hospitality industry. Unless there is once again comfort on the part of the public, restaurants and hotels will be unprofitable. The rule of thumb in NYC is that a drop of 15% in revenues is sufficient to make a restaurant shutter permanently; there has been a large drop in restaurant attendance in Sweden where the state did not enforce lockdowns, owing to a high rate of voluntary self-isolation [17].

The United States (and many other countries worldwide) have spent trillions of dollars on sophisticated weaponry in the past decades, to counter *uncertain* threats. It would be a good idea to question these expenditures first, before doubting the spending to stave off certain pandemics.

Likewise, it would be a good idea to question first the excessive burden on Western economies, particularly the U.S., of measures taken to ensure workplace and transportation safety which, we saw, are driven by the legal system and the tort mechanisms.

Remark 8 (Domain Dependence) *It is not rational to worry about pandemic costs (extremely fat-tailed exposure), while not also questioning other sizable insurance-style expenditures for transportation and workers safety.*

It is therefore incorrect to claim that it is the authorities' response to the pandemic that caused unemployment in the transportation or hospitality industries. As a matter of fact, the arguments proposed by two of the authors [23], last January 2020, were aimed at lowering the economic effect of the pandemic: prevention is orders of magnitude cheaper than the cure– recall that *sed prior est sanitas quam sit curatio morbi*.

We note that many comments of the type "the pandemic has caused *only* 640K fatalities" (as of July 25, 2020) simply ignore the fact that, in practically every location subject to the pandemic, there has been local or governmental action to mitigate it–we do not consider the counterfactual of "what if" because it is not visible.

Remark 9 (Economic Fragility) *The argument in [32] is that we live in an over-optimized environment, in which a slight drop in sales or a change in consumer preferences may cause wild interlocking industry collapses. This nonlinearity is similar to "large a movie theater with a very small door at the times of fire."*

It is more cogent to blame the over-optimized economic structure than the general reaction to the disease.

Early Mitigation and Economic History

We note here that while the Great Plague took place in the fourteenth century, quarantines were enforced five centuries later as economies understood they could not afford recurrences. Between the Habsburg and the Ottoman Empire, there were *lazarettos* along the border, and every active Mediterranean port enforced quarantines for travelers along the expanded silk road, while pilgrim routes were subjected to similar measures. For instance, in the 1830s, in *the Count of Monte Christo*, a traveler from Paris to Ioanina (where Prof. Ioannidis was previously located), had to spend four days in quarantine to get there, while there was no particular threat of disease. In fact, the novelist was underestimating, for historian records show mandatory nine days for ordinary travelers and fifteen days for merchants according to [27, 28]. Economies adapted to early mitigation throughout the centuries preceding our era. Furthermore, the Ottoman Empire has ready *lazaretos* for additional quarantining along specified stations at the first signs of a pandemic.

Mitigation has another effect: to delay and temporize, while we can understand the properties of the disease. While initial treatments are under high opacity, later treatments allow for gains of collective experience[9] .

Selection Bias and Class of Events

In [16], the authors erroneously maintain that choosing tail events as done by [6] is "selection bias". Actually, the standard technique there used is the exact opposite of selection bias: in EVT, one purposely focuses on extremes to derive properties that influence the outcomes, especially from a risk management point of view. One could more reasonably argue that the data in [6] do not contain all the extremes, but, by jackknifing and bootstrapping the data, the authors actually show the robustness of their results to variations and holes in historical observations: the tail index α is consistently lower than 1. In Figure 4 a simple illustration is given, showing that one can be quite radical in dealing with the uncertainty in pandemic fatalities, and still find out that the findings of [6] hold true.

When the authors in [16] state that "Tens of millions of outbreaks with a couple deaths must have happened throughout time," to support their selection bias claim against [6], they seem to overlook the fact that the analysis deals with pandemics and not with a single sternutation. The class of events under considerations in [6] is precisely defined as "pandemics with fatalities in excess of $1K$," and their dataset likely contains most (if not all) of them. Worrying about many missing observations in the left tail of the distribution of pandemic deaths is thus misplaced.

---

[9]Aside from considerations of geronticide, when the costs of the Swedish experiment are finally told, one of the factors will be the early loss of life when later (current) medical practice would have saved them even before a vaccine.

Conditional information

One may be entitled to ask: as we get to know the disease, do the tails get thinner? Early in the game one must rely on conditional information, but as our knowledge of the disease progresses, shouldn't we be allowed to ignore tails?

Alas, no. The scale of the pandemic might change, but the tail properties will remain invariant. Furthermore, there is an additional paradox. If one does not take the pandemic seriously, it will likely run wild (particularly under the connectivity of the modern world, several orders of magnitude higher than in the past [2]). And diseases mutate, increasing or decreasing in both lethality and contagiousness. The argument would therefore resemble the following: "we have not observed many plane crashes lately, let's relax our safety measures".

Finally, we conclude this section with an encouraging point: fat tails do not make the world more complicated and do not cause frivolous worries, to the contrary. Understanding them actually reduces costs of reaction because they tell us what to target–and when to do so. Because network models tend to follow certain patterns to generate large tail events [2, 11], in front of contagious diseases wisdom in action is to kill the exponential growth in the egg via three central measures 1) reducing super-spreader events; 2) monitoring and reducing mobility for those coming from far-away places (via quarantines); 3) looking for cheap measures with large payoffs in terms of the reduction of the multiplicative effects (e.g. face masks[10]). Anything that "demultiplies the multiplicative" helps [35].

Drastic shotgun measures such as lockdowns are the price of avoiding early traveler quarantines and border monitoring; they can be –*temporarily and cum grano salis* –of help, especially in the very early stages of the new contagious disease, when uncertainty is maximal, to help isolating and tracing the infections, and also buying some time for understanding the disease and the way it spreads. Indeed such drastic and painful measures can carry long-lasting damages to the system, not counting an excessive price in terms of personal freedoms.

But they are the price of not having a good coordinated tail risk management in place –to repeat, border monitoring and control of super-spreader events being the very first such measures. And lockdowns are the costs of ignoring arguments such as increased connectivity in our environment and conflating additive and multiplicative risks.

---

[10]Most of the trillions spent could have been saved if authorities understood the double nonlinearities in face masks: 1) the compounding effect of both parties having protection, 2) the nonlinearity of the dose response with disproportional drop in the probability of infection from a reduction in viral load [35].

***

To conclude, as the trader lore transmitted by generations of operators goes, "if you must panic, it pays to panic early."

The Ottoman Empire integrated Byzantine knowledge accumulated since at least the Plague of Justinian; it is sad to see ancient cultures more risk-conscious, better learners from history, and economically more effective than modern governments. They avoided modern "evidence based" reductions that, as we saw, are insulting to both science and wisdom. And, had it not been for such a collective ancestral risk-awareness and understanding of asymmetry, we doubt that many of us would be here today.

Now, what did we learn from the pandemic? That an intelligent application of the precautionary principle [23] consists in formulating decisions that are wise in both foresight and hindsight. Here again, this is ancient: it maps to Aristotle's *phronesis* as presented in his *Nichomachean Ethics*.

Bibliography

[1] M. Ackermann, S.E. Verleden, M. Kuehnel, A. Haverich, T. Welte, F. Laenger, A. Vanstapel, C. Werlein, H. Stark, A. Tzankov, W.W. Li, V.W. Li, S.J. Mentzer, D. Jonigk (2020). Pulmonary Vascular Endothelialitis, Thrombosis, and Angiogenesis in Covid-19. New England Journal of Medicine 383, 120-128.

[2] R. Albert and A.-L. Barabasi (2002). Statistical mechanics of complex networks. Reviews of Modern Physics 74: 47.

[3] M.A. Amaral-Turkman, K.F. Turkman (1990). Optimal alarm systems for autoregressive process; a Bayesian approach. Computational Statistics and Data Analysis 19, 307-314.

[4] P. Cirillo, N.N. Taleb (2016). On the statistical properties and tail risk of violent conflicts. Physica A: Statistical Mechanics and its Applications 452, 29-45.

[5] P. Cirillo, N.N. Taleb (2016). Expected shortfall estimation for apparently infinite-mean models of operational risk. Quantitative Finance 16, 1485-1494.

[6] P. Cirillo, N.N. Taleb (2020). Tail risk of contagious diseases. Nature Physics 16, 606-613.

[7] L. de Haan, A. Ferreira (2006). *Extreme Value Theory: An Introduction*. Springer.

[8] J. de Maré (1980). Optimal prediction of catastrophes with application to Gaussian process. Annals of Probability 8, 841-850.

[9] P. Embrechts, C. Klüppelberg, T. Mikosch (2003). *Modelling Extremal Events*. Springer.

[10] M. Falk, J. Hüsler J, R. D. Reiss R-D (2004). *Laws of small numbers: extremes and rare events*, Birkhäuser.

[11] U. Garibaldi, E. Scalas (2010). Finitary Probabilistic Methods in Econophysics. Cambridge: Cambridge University Press.

[12] A. Gelman (2020). Some forecasting for COVID-19 has failed: a discussion of Taleb and Ioannidis et al.. Available online at https://statmodeling.stat.columbia.edu/2020/06/17/some-forecasting-for-covid-19-has-failed-a-discussion-of-taleb-and-ioannidis-et-al

[13] S.O. Hansson (2004). Fallacies of Risk. Journal of Risk Research 7, 353-360.

[14] J.P.A. Ioannidis (2020). A fiasco in the making? As the coronavirus pandemic takes hold, we are making decisions without reliable data. Stat March 17, https://www.statnews.com/2020/03/17/a-fiasco-in-the-making-as-the-coronavirus-pandemic-takes-hold-we-are-making-decisions-without-reliable-data/.

[15] J.P.A. Ioannidis, C. Axfors, D.G. Contopoulos-Ioannidis (2000a). Population-level COVID-19 mortality risk for non-elderly individuals overall and for non-elderly individuals without underlying diseases in pandemic epicenters. Environmental Research 188, 109890.

[16] J.P.A. Ioannidis, S. Cripps, M.A. Tanner (2020b). Forecasting for COVID-19 has failed. International Institute of Forecasters, available at https://forecasters.org/blog/2020/06/14/forecasting-for-covid-19-has-failed

[17] S. C. L. Kamerlin, Peter M Kasson (2020) Managing COVID-19 spread with voluntary public-health measures: Sweden as a case study for pandemic control, Clinical Infectious Diseases, https://doi.org/10.1093/cid/ciaa864

[18] G. Lindgren (1975). Prediction for a random time point. Annals of Probability 3, 412-433.

[19] E.N. Lorenz (1963). Deterministic non-periodic flow. Journal of the Atmospheric Sciences. 20(2): 130-141.

[20] R. Lucas (1976). Econometric Policy Evaluation: A Critique. In K. Brunner, A. Meltzer (eds.). The Phillips Curve and Labor Markets. Carnegie-Rochester Conference Series on Public Policy 1, 19-46. American Elsevier.

[21] J. Nešlehová, P. Embrechts, V. Chavez-Demoulin (2006). Infinite-mean models and the LDA for operational risk. Journal of Operational Risk 1, 3-25.

[22] J.C. Ngai, F.W. Ko, S..W. To, M. Tong, D.S. Hui DS (2010). The long-term impact of severe acute respiratory syndrome on pulmonary function, exercise capacity and health status. Respirology 15, 543-550.

[23] J. Norman, Y. Bar-Yam, N.N. Taleb (2020). Systemic Risk of Pandemic via Novel Pathogens - Coronavirus: A Note. New England Complex Systems Institute.

[24] O. Peters, M. Gell-Mann (2016). Evaluating gambles using dynamics. Chaos: An Interdisciplinary Journal of Nonlinear Science 26, 023103.

[25] E. M. Rauch and Y. Bar-Yam (2006) Long-range interaction and evolutionary stability in a predator-prey system, Physical Review E 73: 020903.

[26] N.L. Rose (1992). Fear of flying? Economic analysis of airline safety. Journal of Economic Perspectives, 6(2), 75-94.

[27] A Roberts (2017) in Nukhet Varlik (ed.). Plague and Contagion in the Islamic Mediterranean. Arc Humanities Press.

[28] G Sariyildiz and O Daglar Macar (2017) in Nukhet Varlik (ed.). Plague and Contagion in the Islamic Mediterranean. Arc Humanities Press.

[29] K. Søreide, J. Hallet, J.B. Matthews, A.A. Schnitzbauer, P.D. Line, P. Lai, J. Otero, D. Callegaro, S.G. Warner, N.N. Baxter, C. Teh, J. Ng-Kamstra, J.G. Meara, L. Hagander, L. Lorenzon (2020). Immediate and long-term impact of the COVID-19 pandemic on delivery of surgical services. The British Journal of Surgery, https://doi.org/10.1002/bjs.11670.

[30] M. Strathern (1997). Improving ratings: Audit in the British University system. European Review 5, 305-321.

[31] A. Svensson, R. Lindquist, G. Lindgren (1996). Optimal prediction of catastrophes in autoregressive moving average processes. Journal of Time Series Analysis 17, 511-531.

[32] N.N. Taleb (2001-2018). *Incerto: Fooled by Randomness, The Black Swan, The Bed of Procrustes, Antifragile, and Skin in the Game* . Penguin

[33] N.N. Taleb (2020a). *Statistical Consequences of Fat Tails*. STEM Academic Press.

[34] N. N. Taleb (2020b). On the statistical differences between binary forecasts and real-world payoffs. International Journal of Forecasting, in press https://doi.org/10.1016/j.ijforecast.2019.12.004.

[35] N.N. Taleb (2020c) The Masks Masquerade. https://medium.com/incerto/the-masks-masquerade-7de897b517b7

[36] N.N. Taleb, P. Cirillo (2019). The Decline of Violent Conflict: What do the data really say? In A. Toje, N.V.S. Bård, eds. The Causes of Peace: What We Know Now. Nobel Symposium Proceedings. Norwegian Nobel Institute, 57-85.

[37] P.E. Tetlock, D. Gardner (2016). *Superforecasting: The art and science of prediction*. Random House.

[38] R. Viertl (1995). *Statistical Methods for Non-Precise Data*. CRC Press.

[39] M.L. Weitzman (2009). On modeling and intepreting the economics of catastrophic climate change. Review of Economics and Statistics 1, 1-19.

Tail Properties of Contagious Diseases

Pasquale Cirillo[1] and Nassim Nicholas Taleb[2,*]

[1]Applied Probability Group, Delft University of Technology.
[2]Tandon School of Engineering, New York University.
*Corresponding author: nnt1@nyu.edu

Applying a modification of Extreme value Theory (thanks to a dual distribution technique by the authors in [4]) on data over the past 2,500 years, we show that pandemics are extremely fat-tailed in terms of fatalities, with a marked potentially existential risk for humanity. Such a macro property should invite the use of Extreme Value Theory (EVT) rather than naive interpolations and expected averages for risk management purposes. An implication is that potential tail risk overrides conclusions on decisions derived from compartmental epidemiological models and similar approaches.

1 Introduction and Policy Implications

We examine the distribution of fatalities from major pandemics in history (spanning about 2,500 years), and build a statistical picture of their tail properties. Using tools from Extreme Value Theory (EVT), we show for that the distribution of the victims of infectious diseases is extremely fat-tailed, more than what one could be led to believe from the outset[1].

A non-negative continuous random variable X is fat-tailed, in the regular variation class, if its survival function $S(x) = P(X \geq x)$ decays as a power law $x^{-\frac{1}{\xi}}$, the more we move into the tail[2], that is for x growing towards the right endpoint of X. The parameter ξ is known as the tail parameter, and it governs the fatness of the tail (the larger ξ the fatter the tail) and the existence of moments ($E[X^p] < \infty$ if and only if $\xi < 1/p$). In some literature,

---

[1]In this comment we do not discuss the possible generating mechanisms behind these fat tails, a topic of separate research. Networks analysis, e.g. [1], proposes mechanisms for the spreading of contagion and the existence of super spreaders, a plausible joint cause of fat tails. Likewise simple automata processes can lead to high uncertainty of outcomes owing to "computational irreducibility" [28].

[2]More technically, a non-negative continuous random variable X has a fat-tailed distribution (in the maximum domain of attraction of the Fréchet distribution), if its survival function is regularly varying, i.e. $S(x) = L(x)x^{-\frac{1}{\xi}}$, where $L(x)$ is a slowly varying function, such that $\lim_{x \to \infty} \frac{L(cx)}{L(x)} = 1$ for $c > 0$ [7,9].

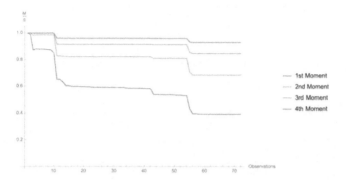

Figure 1: Maximum to Sum plot (MS plot) of the average death numbers in pandemic events in history, as per Table 1.

e.g. [6], the tail index is re-parametrized as $\alpha = 1/\xi$, and its interpretation is naturally reversed.

While it is known that fat tails represent a common–yet often ignored [19] in modeling–regularity in many fields of science and knowledge [6], for the best of our knowledge, only war casualties and operational risk losses show a behavior [4,5,15] as erratic and wild as the one we observe for pandemic fatalities.

The core of the problem is shown in Figure 1, with the Maximum-to-Sum plot [9] of the number of pandemic fatalities in history (data in Table 1). Such a plot relies on a simple consequence of the law of large numbers: for a sequence $X_1, X_2, ..., X_n$ of nonnegative i.i.d. random variables, if $E[X^p] < \infty$ for $p = 1, 2, 3...$, then $R_n^p = M_n^p / S_n^p \to^{a.s.} 0$ as $n \to \infty$, where $S_n^p = \sum_{i=1}^n X_i^p$ is the partial sum of order p, and $M_n^p = \max(X_1^p, ..., X_n^p)$ the corresponding partial maximum. Figure 1 clearly shows that no finite moment is likely to exist for the number of victims in pandemics, as the R_n ratio does not converge to 0 for $p = 1, 2, 3, 4$, no matter how many data points we use. Such a behavior hints that the victims distribution has such a fat right tail that not even the first theoretical moment is finite. We are looking at a phenomenon for which observed quantities such as the naive sample average and standard deviation are therefore meaningless for inference.

However, Figure 1 (or a naive use of EVT) does not imply that pandemic risk is actually infinite and there is nothing we can do or model. Using the methodology we developed to study war casualties [4,20], we are in fact able to extract useful information from the data, quantifying the large yet finite risk of pandemic diseases. The method provides in fact rough estimates for quantities not immediately observable in the data.

The tail wags the dog effect

Centrally, the more fat-tailed the distribution, the more "the tail wags the dog", that is, the more statistical information resides in the extremes and the less in the "bulk" (that is the events of high frequency), where it becomes almost noise. This makes EVT the most effective approach, and our sample of extremes very highly sufficient and informative for risk management purposes[3].

The fat-tailedness of the distribution of pandemic fatalities has the following policy implications, useful in the wake of the Covid-19 pandemic.

First, it should be evident that one cannot compare fatalities from multiplicative infectious diseases (fat-tailed, like a Pareto) to those from car accidents, heart attacks or falls from ladders (thin-tailed, like a Gaussian). Yet this is a common (and costly) error in policy making, and in both the decision science and the journalistic literature[4]. Some research papers even criticise people's "paranoïa" with respect to pandemics, not understanding that such a paranoïa is merely responsible (and realistic) risk management in front of potentially destructive events [19]. The main problem is that those articles–often relied upon for policy making –consistently use the wrong thin-tailed distributions, underestimating tail risk, so that every conservative or preventative reaction is bound to be considered an overreaction.

Second, epidemiological models like the SIR [13] differential equations, sometimes supplemented with simulation experiments like [11], while useful for scientific discussions for the bulk of the distributions of infections and deaths, or to understand the dynamics of events after they happened, should never be used for precautionary risk management, which should focus on maxima and tail exposures instead. It is highly unrigorous to use naive (and reassuring) statistics, like the expected average outcome of compartmental models, or one or more point estimates, as a motivation for policies. Owing to the compounding effect of parameters' uncertainty, the "tail wagging the dog" effect easily invalidates both point estimates and scenario analyses[5].

EVT is the natural candidate to handle pandemics. It was born to cope with maxima [10], and it evolved to deal with tail risk in a robust way, even with a limited number of observations and the uncertainty associated with it [9]. In the Netherlands, for example, EVT was used to get a handle on the distribution of the maxima–not the average!–of sea levels in order to build dams and dykes high and strong enough for the safety of citizens [7].

---

[3]Since the law of large numbers works slowly under fat tails, the bulk becomes increasingly dominated by noise, and averages and higher moments–even when they exist–become uninformative and unreliable, while extremes are rich in information [19].

[4]Sadly, this mistake is sometimes made by professional statisticians as well. Thin tailed (discrete) variables are subjected to Chernov bounds, unlike fat-tailed ones [19].

[5]The current Covid-19 pandemic is generating a lot of research, and finally some scholars are looking at the impact of parameters' uncertainty on the scenarios generated by epidemiological models, e.g. [8].

Finally, EVT-based risk management is compatible with the (non-naïve) precautionary principle of [16], which should be the leading driver for policy decisions under jointly systemic and extreme risks.

2 Data and descriptive statistics

We investigate the distribution of deaths from the major epidemic and pandemic diseases of history, from 429 BC until now. The data are available in Table 1, together with their sources, and only refer to events with more than 1K estimated victims, for a total of 72 observations. As a consequence, potentially high-risk diseases, like the Middle East Respiratory Syndrome (MERS), do not appear in our collection[6]. All diseases whose end year is 2020 are to be taken as still occurring worldwide, as for the running COVID-19 pandemic.

Three estimates of the reported cumulative death toll have been used: minimum, average and maximum. When the three numbers coincide in Table 1, our sources simply do not provide intervals for the estimates. Since we are well aware of the volatility and possible unreliability of historical data [18, 20], in Section 4 we deal with such an issue by perturbing and omitting observations.

In order to compare fatalities with respect to the coeval population (that is, the relative impact of pandemics), column *Rescaled* of Table 1 provides the rescaled version of column *Avg Est*, using the information in column *Population*[7] [12, 14, 21]. For example, the Antonine plague of 165-180 killed an average of 7.5M people, that is to say 3.7% of the coeval world population of 202M people. Using today's population, such a number would correspond to about 283M deaths, a terrible hecatomb, killing more people than WW2.

For space considerations, we restrict our attention to the actual average estimates in Table 1, but all our findings and conclusions hold true for the lower, the upper and the rescaled estimates as well[8].

Figure 2 shows the histogram of the actual average numbers of deaths in the 72 large contagious events. The distributions appears highly skewed and possibly fat-tailed. The numbers are as follows: the sample average is 4.9M, while the median is 76K, compatibly with the skewness observable in Figure 2. The 90% quantile is 6.5M and the 99% quantile is 137.5M. The sample standard deviation is 19M.

---

[6]Up to the present, MERS has killed 858 people as reported in https://www.who.int/emergencies/mers-cov/en. For SARS the death toll is between 774 and 916 victims until now https://www.nytimes.com/2003/10/05/world/taiwan-revises-data-on-sars-total-toll-drops.html.

[7]Population estimates are by definitions estimates, and different sources can give different results (most of the times differences are minor), especially for the past. However our methodology is robust to this type of variability, as we stress later in the paper.

[8]The differences in the estimates do not change the main message: we are dealing with an extremely erratic phenomenon, characterised by very fat tails.

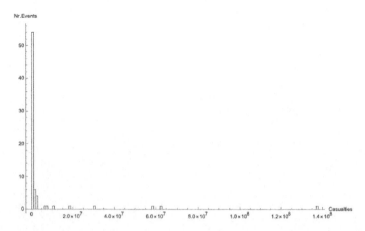

Figure 2: Histogram of the average number of deaths in the 72 contagious diseases of Table 1.

Using common graphical tools for fat tails [9], in Figure 3 we show the log log plot (also known as Zipf plot) of the empirical survival functions for the average victims over the diverse contagious events. In such a plot possible fat tails can be identified in the presence of a linearly decreasing behavior of the plotted curve. To improve interpretability a naive linear fit is also proposed. Figure 3 suggests the presence of fat tails.

The Zipf plot shows a necessary but not sufficient condition for fat-tails [3]. Therefore, in Figure 4 we complement the analysis with a mean excess function plot, or meplot. If a random variable X is possibly fat-tailed, its mean excess function $e_X(u) = E[X - u | X \geq u]$ should grow linearly in the threshold u, at least above a certain value identifying the actual power law tail [9]. In a meplot, where the empirical $e_X(u)$ is plotted against the different values of u, one thus looks for some (more or less) linearly increasing trend, as the one we observe in Figure 4.

A useful tool for the analysis of tails–when one suspects them to be fat–is the nonparametric Hill estimator [9]. For a collection $X_1, ..., X_n$, let $X_{n,n} \leq ... \leq X_{1,n}$ be the corresponding order statistics. Then we can estimate the tail parameter ξ as

$$\hat{\xi} = \frac{1}{k} \sum_{i=1}^{k} \log(X_{i,n}) - \log(X_{k,n}), \qquad 2 \leq k \leq n.$$

In Figure 5, $\hat{\xi}$ is plotted against different values of k, creating the so-called Hill plot [9]. The plot suggests $\xi > 1$, in line with Figure 1, further support-ing the evidence of infinite moments.

Other graphical tools could be used and they would all confirm the point: we are in the presence of fat tails in the distribution of the victims

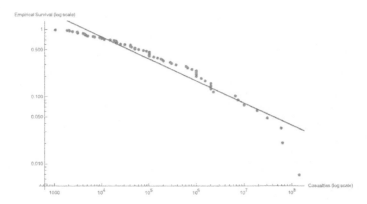

Figure 3: Log log plot of the empirical survival function (Zipf plot) of the actual average death numbers in Table 1. The red line represents a naive linear fit of the decaying tail.

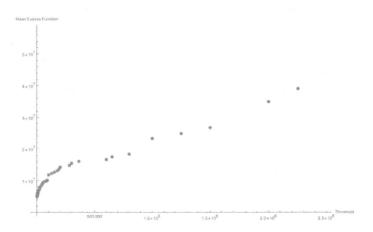

Figure 4: Mean excess function plot (meplot) of the average death numbers in Table 1. The plot excludes 3 points on the top right corner, consistently with the suggestions in [9] about the exclusion of the more volatile observations.

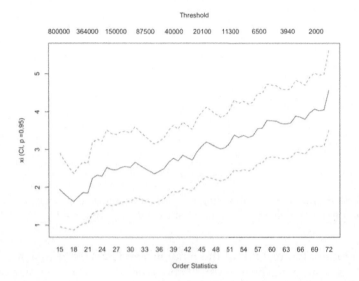

Figure 5: Hill plot of the average death numbers in Table 1, with 95% confidence intervals. Clearly $\xi > 1$, suggesting the non-existence of moments.

of pandemic diseases. Even more, a distribution with possibly no finite moment.

The dual distribution

As we observed for war casualties [4], the non-existence of moments for the distribution of pandemic victims is questionable. Since the distribution of victims is naturally bounded by the coeval world population, no disease can kill more people than those living on the planet at a given time time. We are indeed looking at an *apparently* infinite-mean phenomenon, like in the case of war casualties [4, 20] and operational risk [5].

Let $[L, H]$ be the support of the distribution of pandemic victims today, with $L \gg 0$ to ignore small events not officially definable as pandemic [24]. For what concerns H, its value cannot be larger than the world population, i.e. 7.7 billion people in 2020[9]. Evidently H is so large that the probability of observing values in its vicinity is in practice zero, and one always finds observations below a given $M \ll H < \infty$ (something like 150M deaths using actual data). Thus one could be fooled by data into ignoring H and taking it as infinite, up to the point of believing in an infinite mean phenomenon, as Figure 1 suggests. However notice that a finite upper bound H–no matter how large it is–is not compatible with infinite moments, hence Figure 1 risks to be dangerously misleading.

---

[9]Today's world population [21] can be safely taken as the upper bound also for the past.

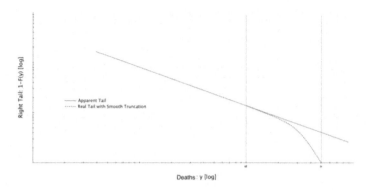

Figure 6: Graphical representation (log-log plot) of what may happen if one ignores the existence of the finite upper bound H, since only M is observed.

In Figure 6, the real tail of the random variable Y with remote upper bound H is represented by the dashed line. If one only observes values up to $M \ll H$, and more or less consciously ignores the existence of H, one could be fooled by the data into believing that the tail is actually the continuous one, the so-called apparent tail [5]. The tails are indeed indistinguishable for most cases, virtually in all finite samples, as the divergence is only clear in the vicinity of H. A bounded tail with very large upper limit is therefore mistakenly taken for an unbounded one, and no model will be able to see the difference, even if epistemologically we are in two extremely different situations. This is the typical case in which critical reasoning, and the a priori analysis of the characteristics of the phenomenon under scrutiny, should precede any instinctive and uncritical fitting of the data.

A solution is the approach of [4,5], which introduces the concept of dual data via a special log-transformation [10]. The basic idea is to find a way of matching naive extrapolations (apparently infinite moments) with correct modelling.

Let L and H be respectively the finite lower and upper bounds of a random variable Y, and define the function

$$\varphi(Y) = L - H \log\left(\frac{H-Y}{H-L}\right). \tag{1}$$

We can easily check that

1. $\varphi \in C^\infty$,

2. $\varphi^{-1}(\infty) = H$,

---

[10]Other log-transformations have been proposed in the literature, but they are all meant to thin the tails, without actually taking care of the upper bound problem: the number of victims can still be infinite. The rationale behind those transformations is given by the observation that if X is a random variable whose distribution function is in the domain of attraction of a Fréchet, the family of fat-tailed distributions, then $\log(X)$ is in the domain of attraction of a Gumbel, the more reassuring family of normals and exponentials [9].

3. $\varphi^{-1}(L) = \varphi(L) = L$.

Then $Z = \varphi(Y)$ defines a new random variable with lower bound L and an infinite upper bound. Notice that the transformation induced by $\varphi(\cdot)$ does not depend on any of the parameters of the distribution of Y, and that $\varphi(\cdot)$ is monotone. From now on, we call the distributions of Y and Z, respectively the real and the dual distribution. It is easy to verify that for values smaller than $M \ll H$, Y and Z are in practice indistinguishable (and do are their quantiles [5]).

As per [4,5], we take the observations in the column "Avg Est" of Table 1, our Y's, and transform them into their dual Z's. We then study the actually unbounded duals using EVT (see Section 3), to find out that the naive observation of infinite moments can makes sense in such a framework (but not for the bounded world population!). Finally, by reverting to the real distribution, we compute the so-called *shadow* means [5] of pandemics, equal to

$$E[Y] = (H - L)e^{\frac{1}{2}\frac{\sigma}{H}} \left(\frac{\sigma}{H\xi}\right)^{\frac{1}{\xi}} \Gamma\left(1 - \frac{1}{\xi}, \frac{\sigma}{H\xi}\right) + L, \tag{2}$$

where $\Gamma(\cdot, \cdot)$ is the gamma function.

Notice that the random quantity Y is defined above L, therefore its expectation corresponds to a tail expectation with respect to the random variable Z, an expected shortfall in the financial jargon, being only valid in the tail above μ [4]. All moments of the random variable Y are called shadow moments in [5], as they are not immediately visible from the data, but from plug-in estimation.

3   The dual tail via EVT and the shadow mean

Take the dual random variable Z whose distribution function G is unknown, and let $z_G = \sup\{z \in \mathbb{R} : G(z) < 1\}$ be its right-end point, which can be finite or infinite. Given a threshold $u < z_G$, we can define the exceedance distribution of Z as

$$G_u(z) = P(Z \le z | Z > u) = \frac{G(z) - G(u)}{1 - G(u)}, \tag{3}$$

for $z \ge u$.

For a large class of distributions G, and high thresholds $u \to z_G$, G_u can be approximated by a Generalized Pareto distribution (GPD) [7], i.e.

$$G_u(z) \approx GPD(z; \xi, \beta, u) = \begin{cases} 1 - (1 + \xi\frac{z-u}{\beta})^{-1/\xi} & \xi \ne 0 \\ 1 - e^{-\frac{z-u}{\beta}} & \xi = 0 \end{cases}, \tag{4}$$

where $z \ge u$ for $\xi \ge 0$, $u \le z \le u - \beta/\xi$ for $\xi < 0$, $u \in \mathbb{R}$, $\xi \in \mathbb{R}$ and $\beta > 0$.

Let us just consider $\xi > 0$, being $\xi = 0$ not relevant for fat tails. From equation (3), we see that $G(z) = (1 - G(u))G_u(z) + G(u)$, hence we obtain

$$G(z) \approx (1 - G(u))GPD(z; \xi, \beta, u) + G(u)$$

$$= 1 - \bar{G}(u)\left(1 + \xi\frac{z - u}{\beta}\right)^{-1/\xi},$$

with $\bar{G}(x) = 1 - G(x)$. The tail of Z is therefore

$$\bar{G}(z) = \bar{G}(u)\left(1 + \xi\frac{z - u}{\beta}\right)^{-1/\xi}. \tag{5}$$

Equation (5) is called the tail estimator of $G(z)$ for $z \geq u$. Given that G is in principle unknown, one usually substitutes $G(u)$ with its empirical estimator n_u/n, where n is the total number of observations in the sample, and n_u is the number of exceedances above u.

Equation (5) then changes into

$$\bar{G}(z) = \frac{n_u}{n}\left(1 + \xi\frac{z - u}{\beta}\right)^{-1/\xi} \approx 1 - GPD(z^*; \xi, \sigma, \mu), \tag{6}$$

where $\sigma = \beta\left(\frac{n_u}{n}\right)^{\xi}$, $\mu = u - \frac{\beta}{\xi}\left(1 - \left(\frac{n_u}{n}\right)^{\xi}\right)$, and $z^* \geq \mu$ is an auxiliary random variable. Both σ and μ can be estimated semi-parametrically, starting from the estimates of ξ and β in equation (4). If $\xi > -1/2$, the preferred estimation method is maximum likelihood [7], while for $\xi \leq -1/2$ other approaches are better used [9]. For both the exceedances distribution and the recovered tail, the parameter ξ is the same, and it also coincides with the tail parameter we have used to define fat tails[11].

One can thus study the tail of Z without caring too much about the rest of the distribution, i.e. the part below u. All in all, the most destructive risks come from the right tail, and not from the first quantiles or even the bulk of the distribution. The identification of the correct u is a relevant question in extreme value statistics [7,9]. One can rely on heuristic graphical tools [3], like the Zipf plot and the meplot we have seen before, or on statistical tests for extreme value conditions [10] and GPD goodness-of-fit [2].

What is important to stress–once again–is that the GPD fit needs to be performed on the dual quantities, to be statistically and epistemologically correct. One could in fact work with the raw observation directly, without the log-transformation of Equation (1), surely ending up with $\xi > 1$, in line with Figures 1 and 5. But a similar approach would be wrong and naive, because only the dual observations are actually unbounded.

---

[11]Moreover, when maximum likelihood is used, the estimate of ξ would correspond to $1/\alpha$, where α is estimated according to [6].

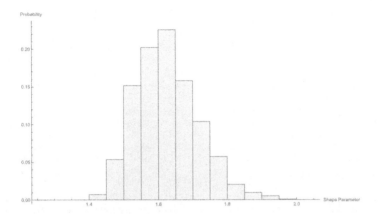

Figure 7: Values of the shape parameter ξ over 10,000 distorted copies of the the dual versions of the average deaths in Table 1, allowing for a random variation of ±20% for each single observation. The ξ parameter consistently indicates an apparently infinite-mean phenomenon.

Working with the dual observations, we find out that the best GPD fit threshold is around 200K victims, with 34.7% of the observations lying above. For what concerns the GPD parameters, we estimate ξ = 1.62 (standard error 0.52), and β = 1'174.7K (standard error 536.5K). As expected ξ > 1 once again supporting the idea of an infinite first moment[12]. Visual inspections and statistical tests [2, 10] support the goodness-of-fit for the exceedance distribution and the tail.

Given ξ and β, we can use Equations (2) and (6) to compute the shadow mean of the numbers of victims in pandemics. For actual data we get a shadow mean of 20.1M, which is definitely larger (almost 1.5 times) than the corresponding sample tail mean of 13.9M (this is the mean of all the actual numbers above the 200K threshold.). Combining the shadow mean with the sample mean below the 200K threshold, we get an overall mean of 7M instead of the naive 4.9M we have computed initially. It is therefore important to stress that a naive use of the sample mean would induce an underestimation of risk, and would also be statistically incorrect.

4 Data reliability issues

As observed in [4, 18, 20] for war casualties, but the same reasoning applies to pandemics of the past, the estimates of the number of victims are not at

---

[12]Looking at the standard error of ξ, one could argue that, with more data from the upper tail, the first moment could possibly become finite, yet there would be no discussion about the non existence of the second moment, and thus the unreliability of the sample mean [19]. Pandemic fatalities would still be an extremely erratic phenomenon, with substantial tail risk in the number of fatalities. In any case, Figures 1 and 5 make us prefer to consider the first moment as infinite, and not to trust sample averages.

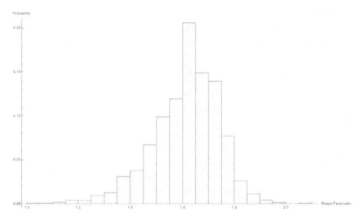

Figure 8: Values of the shape parameter ξ over 10,000 jackknifed versions of the dual versions of the actual average numbers in Table 1, when allowing at least 1% and up to about 10% of the observations to be missing. The ξ parameter consistently indicates an apparently infinite-mean phenomenon.

all unique and precise. Figures are very often anecdotal, based on citations and vague reports, and usually dependent on the source of the estimate. In Table 1, it is evident that some events vary considerably in estimates.

Natural questions thus arise: are the tail risk estimates of Section 3 robust? What happens if some of the casualties estimates change? What is the impact of ignoring some events in our collection? The use of extreme value statistics in studying tail risk already guarantees the robustness of our estimates to changes in the underlying data, when these lie below the threshold u. However, to verify robustness more rigorously and thoroughly, we have decided to stress the data, to study how the tails potentially vary.

First of all, we have generated 10K distorted copies of our dual data. Each copy contains exactly the same number of observations as per Table 1, but every data point has been allowed to vary between 80% and 120% of its recorded value before imposing the log-transformation of Equation (1). In other words, each of the 10K new samples contains 72 observations, and each observation is a (dual) perturbation ($\pm 20\%$) of the corresponding observation in Table 1.

Figure 7 contains the histogram of the ξ parameter over the 10K distorted copies of the dual numbers. The values are always above 1, indicating an apparently infinite mean, and the average value is 1.62 (standard deviation 0.10), in line with our previous findings. Our tail estimates are thus robust to imprecise observations. Consistent results hold for the β parameter.

But it also true that our data set is likely to be incomplete, not containing all epidemics and pandemics with more than 1K victims, or that some of the events we have collected are too biased to be reliable and should be discarded anyway. To account for this, we have once again generated 10K

Name	Start Year	End Year	Lower Est (1k)	Avg Est (1k)	Upper Est (1k)	Rescaled Avg Est (1k)	Population (1mio)	Source
Plague of Athens	-429	-426	75	88	100	13376	50	[24]
Antonine Plague	165	180	5000	7500	10000	283355	202	[24]
Plague of Cyprian	250	266	1000	1000	1000	37227	205	[24]
Plague of Justinian	541	542	25000	62500	100000	2246550	213	[24]
Plague of Amida	562	562	30	30	30	1078	213	[21]
Roman Plague of 590	590	590	10	20	30	719	213	[24]
Plague of Sheroe	627	628	100	100	100	3594	213	[26]
Plague of the British Isles	664	689	150	175	200	6290	213	[24]
Plague of Basra	688	689	200	200	200	7189	213	[21]
Japanese smallpox epidemic	735	737	2000	2000	2000	67890	226	[24]
Black Death	1331	1353	75000	137500	200000	2678283	392	[24]
Sweating sickness	1485	1551	10	10	10	166	461	[24]
Smallpox Epidemic in Mexico	1520	1520	5000	6500	8000	107684	461	[24]
Cocolitzli Epidemic of 1545–1548	1545	1548	5000	10000	15000	165668	461	[24]
1563 London plague	1562	1564	20	20	20	277	554	[24]
Cocoliztli epidemic of 1576	1576	1580	2000	2250	2500	31045	554	[24]
1592–93 London plague	1592	1593	20	20	20	275	554	[24]
Malta plague epidemic	1592	1593	3	3	3	41	554	[24]
Plague in Spain	1596	1602	600	650	700	8969	554	[24]
New England epidemic	1616	1620	7	7	7	97	554	[24]
Italian plague of 1629–1631	1629	1631	280	280	280	3863	554	[24]
Great Plague of Sevilla	1647	1652	150	150	150	2070	554	[24]
Plague in Kingdom of Naples	1656	1658	1250	1250	1250	15840	603	[16]
Plague in the Netherlands	1663	1664	24	24	24	306	603	[24]
Great Plague of London	1665	1666	100	100	100	1267	603	[24]
Plague in France	1668	1668	40	40	40	507	603	[24]
Malta plague epidemic	1675	1676	11	11	11	143	603	[24]
Great Plague of Vienna	1679	1679	76	76	76	963	603	[24]
Great Northern War plague outbreak	1700	1721	176	192	208	2427	603	[25]
Great Smallpox Epidemic in Iceland	1707	1709	18	18	18	228	603	[24]
Great Plague of Marseille	1720	1722	100	100	100	1267	603	[24]
Great Plague of 1738	1738	1738	50	50	50	470	814	[24]
Russian plague of 1770–1772	1770	1772	50	50	50	470	814	[24]
Persian Plague	1772	1772	2000	2000	2000	15444	990	[24]
Ottoman Plague Epidemic	1812	1819	300	300	300	2317	990	[26]
Caragea's plague	1813	1813	60	60	60	463	990	[26]
Malta plague epidemic	1813	1814	5	5	5	35	990	[24]
First cholera pandemic	1816	1826	100	100	100	772	990	[24]
Second cholera pandemic	1829	1851	100	100	100	772	990	[24]
Typhus epidemic in Canada	1847	1848	20	20	20	154	990	[24]
Third cholera pandemic	1852	1860	1000	1000	1000	6053	1263	[24]
Cholera epidemic of Copenhagen	1853	1853	5	5	5	29	1263	[26]
Third plague pandemic	1855	1960	15000	18500	22000	111986	1263	[24],[26]
Smallpox in British Columbia	1862	1863	3	3	3	18	1263	[26]
Fourth cholera pandemic	1863	1875	600	600	600	3632	1263	[26]
Fiji Measles outbreak	1875	1875	40	40	40	242	1263	[26]
Yellow Fever	1880	1900	100	125	150	757	1263	[22]
Fifth cholera pandemic	1881	1896	9	9	9	42	1654	[24]
Smallpox in Montreal	1885	1885	3	3	3	14	1654	[24]
Russian flu	1889	1890	1000	1000	1000	4620	1654	[24]
Sixth cholera pandemic	1899	1923	800	800	800	3696	1654	[26]
China plague	1910	1912	40	40	40	185	1654	[24]
Encephalitis lethargica pandemic	1915	1926	1500	1500	1500	6930	1654	[24]
American polio epidemic	1916	1916	6	7	7	30	1654	[24]
Spanish flu	1918	1920	17000	58500	100000	193789	2307	[24]
HIV/AIDS pandemic	1920	2020	25000	30000	35000	61768	3712	[22]
Poliomyelitis in USA	1946	1946	2	2	2	5	2948	[24]
Asian flu	1957	1958	2000	2000	2000	5186	2948	[24]
Hong Kong flu	1968	1969	1000	1000	1000	2102	3637	[24]
London flu	1972	1973	1	1	1	2	3866	[24]
Smallpox epidemic of India	1974	1974	15	15	15	29	4016	[24]
Zimbabwean cholera outbreak	2008	2009	4	4	4	5	6788	[24]
Swine Flu	2009	2009	152	364	575	409	6788	[24]
Haiti cholera outbreak	2010	2020	10	10	10	11	7253	[24]
Measles in D.R. Congo	2011	1018	5	5	5	5	7253	[24]
Ebola in West Africa	2013	2016	11	11	11	12	7176	[24]
Indian swine flu outbreak	2015	2015	2	2	2	2	7253	[24]
Yemen cholera outbreak	2016	2020	4	4	4	4	7643	[24]
2018–19 Kivu Ebola epidemic	2018	2020	2	2	3	2	7643	[24]
2019-20 COVID-19 Pandemic	2019	2020	117	133.5	150	50	7643	[23]
Measles in D.R. Congo	2019	2020	5	5	5	5	7643	[26]
Dengue fever	2019	2020	2	2	2	2	7643	[24]

Table 1: The data set used for the analysis. All estimates in thousands, apart from coeval population, which is expressed in millions. For Covid-19 [24], the upper estimate includes the supposed number of Chinese victims (42K) for some Western media.

copies of our sample via jackknife. Each new dual sample is obtained by removing from 1 to 7 observations at random, so that one sample could not contain the Spanish flu, while another could ignore the Yellow Fever and AIDS. In Figure 8 we show the impact of such a procedure on the ξ parameter. Once again, the main message of this work remains unchanged: we are looking at a very fat-tailed phenomenon, with an extremely large tail risk and potentially destructive consequences, which should not be downplayed in any serious policy discussion.

Bibliography

[1] R. Albert and A.-L. Barabasi (2002). Statistical mechanics of complex networks. Reviews of Modern Physics 74: 47.

[2] M. Arshad , M.T. Rasool, M.I. Ahmad (2003). Anderson Darling and Modified Anderson Darling Tests for Generalized Pareto Distribution. Journal of Applied Sciences 3, 85-88.

[3] P. Cirillo (2013). Are your data really Pareto distributed? *Physica A: Statistical Mechanics and its Applications* 392, 5947-5962.

[4] P. Cirillo, N.N. Taleb (2016). On the statistical properties and tail risk of violent conflicts. Physica A: Statistical Mechanics and its Applications 452, 29-45.

[5] P. Cirillo, N.N. Taleb (2016). Expected shortfall estimation for apparently infinite-mean models of operational risk. Quantitative Finance 16, 1485-1494.

[6] A. Clauset, C.R. Shalizi, M.E.J. Newman (2009). Power-law distributions in empirical data. SIAM Review 51, 661-703.

[7] L. de Haan, A. Ferreira (2006). *Extreme Value Theory: An Introduction.* Springer.

[8] C. Donnat, S. Holmes (2020). Modeling the heterogeneity in COVID-19 as reproductive number and its impact on predictive scenarios. arXiv:2004.05272.

[9] P. Embrechts, C. Klüppelberg, T. Mikosch (2003). *Modelling Extremal Events.* Springer.

[10] M. Falk, J. Hüsler J, R. D. Reiss R-D (2004). *Laws of small numbers: extremes and rare events*, Birkhäuser.

[11] N. Ferguson, D. Laydon, G. Nedjati-Gilani et alii (2020). Report 9: Impact of non-pharmaceutical interventions (NPIs) to reduce COVID19 mortality and healthcare demand. Available online at https://www.imperial.ac.uk/media/imperial-college/medicine/mrc-gida/2020-03-16-COVID19-Report-9.pdf

[12] K. Goldewijk, K. Beusen, P. Janssen (2010). Long term dynamic modeling of global population and built-up area in a spatially explicit way, hyde 3.1. *The Holocene* 20, 565-573.

[13] H.W. Hethcote (2000). The mathematics of infectious diseases. *SIAM review* 42, 599-653.

[14] K. Klein Goldewijk, G. van Drecht (2006). HYDE 3.1: Current and historical population and land cover. In A. F. Bouwman, T. Kram, K. Klein Goldewijk. *Integrated modelling of global environmental change. An overview of IMAGE 2.4.* Netherlands Environmental Assessment Agency.

[15] J. Nešlehová, P. Embrechts, V. Chavez-Demoulin (2006). Infinite-mean models and the LDA for operational risk. Journal of Operational Risk 1, 3-25.

[16] J.Norman, Y. Bar-Yam, N.N. Taleb (2020). Systemic Risk of Pandemic via Novel Pathogens - Coronavirus: A Note. New England Complex Systems Institute.

[17] S. Scasciamacchia, L. Serrecchia, L. Giangrossi, G. Garofolo, A. Balestrucci, G. Sammartino (2012). Plague Epidemic in the Kingdom of Naples, 1656â1658. Emerging Infectious Diseases 18, 186-188.

[18] T. B. Seybolt, J. D. Aronson, B. Fischhoff, eds. (2013). *Counting Civilian Casualties, An Introduction to Recording and Estimating Nonmilitary Deaths in Conflict.* Oxford University Press.

[19] N.N. Taleb (2020). *Statistical Consequences of Fat Tails.* STEM Academic Press.

[20] N.N. Taleb, P. Cirillo (2019). The Decline of Violent Conflict: What do the data really say? In A. Toje, N.V.S. Bård, eds. The Causes of Peace: What We Know Now. Nobel Symposium Proceedings. Norwegian Nobel Institute, 57-85.

[21] United Nations - Department of Economic and Social Affairs (2015). *2015 Revision of World Population Prospects.* UN Press.

[22] Ancient History Encyclopedia, retrieved on March 30, 2020: `https://www.ancient.eu/article/1528/plague-in-the-ancient--medieval-world/`.

[23] Visual Capitalist, Visualizing the History of Pandemics, retrieved on March 30, 2020: `https://www.visualcapitalist.com/history-of-pandemics-deadliest/`.

[24] World Health Organization, Covid-19 page, retrieved on April 13, 2020: `https://www.who.int/emergencies/diseases/novel-coronavirus-2019`

[25] Wikipedia List of epidemics, retrieved on March 30, 2020: `https://en.wikipedia.org/wiki/List_of_epidemics`.

[26] Wikipedia Great Northern War plague outbreak, retrieved on March 30, 2020: `https://en.wikipedia.org/wiki/Great_Northern_War_plague_outbreak`.

[27] Weblist of Epidemics Compared to Coronavirus, retrieved on March 30, 2020: `https://listfist.com/list-of-epidemics-compared-to-coronavirus-covid-19`.

[28] S. Wolfram (2002). *A New Kind of Science.* Wolfram Media.

Modelling SARS-CoV-2 Coevolution with Genetic Algorithms

Aymeric Vié*

Mathematical Institute, University of Oxford
Institute of New Economic Thinking, University of Oxford
*Corresponding author: vie@maths.ox.ac.uk

At the end of 2020, policy responses to the SARS-CoV-2 outbreak have been shaken by the emergence of virus variants, impacting public health and policy measures worldwide. The emergence of these strains suspected to be more contagious, more severe, or even resistant to antibodies and vaccines, seem to have taken by surprise health services and policymakers, struggling to adapt to the new variants constraints. Anticipating the emergence of these mutations to plan ahead adequate policies, and understanding how human behaviors may affect the evolution of viruses by coevolution, are key challenges. In this article, we propose coevolution with genetic algorithms (GAs) as a credible approach to model this relationship, highlighting its implications, potential and challenges. Because of their qualities of exploration of large spaces of possible solutions, capacity to generate novelty, and natural genetic focus, GAs are relevant for this issue. We present a dual GA model in which both viruses aiming for survival and policy measures aiming at minimising infection rates in the population, competitively evolve. This artificial coevolution system may offer us a laboratory to "debug" our current policy measures, identify the weaknesses of our current strategies, and anticipate the evolution of the virus to plan ahead relevant policies. It also constitutes a decisive opportunity to develop new genetic algorithms capable of simulating much more complex objects. We highlight some structural innovations for GAs for that virus evolution context that may carry promising developments in evolutionary computation, artificial life and AI.

1   Introduction

As early as June 2020, the initial SARS-CoV-2 strain identified in China was replaced as the dominant variant by the D614G mutation (Figure 1). Appeared in January 2020, this strain differed because of a substitution in the gene encoding the spike protein. The D614G substitution has been

Figure 1: Shift over time from orange (the original D type of the virus) to blue (the now-widespread G form, D614G); (Los Alamos National Laboratory, 2020)

found to have increased infectivity and transmission (WHO, 2020a; Korber et al., 2020).

On November 5 2020, a new strain of SARS-CoV-2 was reported in Denmark (WHO, 2020b), linked with the mink industry. The "unique" mutations identified in one cluster, "Cluster 5", seemingly as contagious or severe as others, has been found to moderately decrease the sensitivity of the disease to neutralising antibodies. Culling of farmed minks, increase of genome sequencing activities and numerous closing of borders to Denmark residents, followed.

On 14 December 2020, the United Kingdom reported a new variant VOC 202012/01, with a remarkable number of 23 mutations, with unclear origin (Kupferschmidt, 2020). Early analyses have found that the variant has increased transmissibility, though no change in disease severity was identified (WHO, 2020a). One of these 23 mutations, the deletion at position 69/70del, was found to affect the performance of some PCR tests, currently at the center of national testing strategies. Quickly becoming dominant, this variant was held responsible for a significant increase in mortality, ICU occupation and infections across the country (Iacobucci, 2021; Wallace and Ackland, 2021).

On 18 December, the variant 501Y.V2 was detected in South Africa, after rapidly displacing other virus lineages in the region. Preliminary studies showed that this variant was associated with a higher viral load, which may cause increased transmissibility (WHO, 2020a). Recent findings have shown that this variant significantly reduced the efficacy of vaccines (Mahase, 2021).

RNA viruses have high mutation rates (Duffy, 2018). Although many mutations are not beneficial for the organisms, and some are inconsequential, some small fraction of them are beneficial. We refer the reader to (Duffy, 2018) and Domingo et al. (1996) for a discussion on RNA viruses mutation rates. The consequences of these high mutation rates notably are higher evolvability, i.e. higher capacity to adapt to changing environments. This allows them to emerge in new hosts, escape vaccine-induced immunity, or circumvent disease resistance. However, RNA viruses seem to be just below the threshold for critical error: if the majority of mutations are deleterious, higher mutation rates may cause ecological collapse in the virus population. As a RNA virus (Lima, 2020), SARS-CoV-2 shares these characteristics, and mutates very frequently (Phan, 2020; Benvenuto et al. (2020), Matyásek, and Kovarík, 2020). Especially relevant for this class of virus, the priorities of many researchers including the WHO Virus Evolution Working Group, have been to strengthen ways to identify relevant mutations, study their characteristics and impacts, as well as outlining mitigation strategies to respond to these mutations (WHO, 2020a).

Anticipating the emergence of these mutations to plan ahead adequate policies, and understanding how human behaviors may affect the evolution of viruses by coevolution, are key challenges. Human adaptation of policies and behaviors can impact the reproduction of SARS-CoV-2, and target specific characteristics such as airborne transmission. The impact of human policies and behaviors on outbreak trajectory, the evaluation of non pharmaceutical measures, have been the object of numerous analyses. However, most of these analyses do not include the possibility for viruses to mutate, with novel effects and increased transmission rates. The space of possible virus strains is huge and to some extent quasi open-ended, challenging modelling attempts of this arms' race.

In this article, we propose coevolution with genetic algorithms (GAs) as a credible approach to model this relationship, highlighting its implications, potential and challenges. We provide a proof of concept-implementation of this coevolution dual-GA. Because of their qualities of exploration of large spaces of possible solutions, capacity to generate novelty, and natural genetic focus, GAs are relevant for this issue. We present a dual GA model in which both viruses aiming for survival and policy measures aiming at minimising infection rates in the population, competitively evolve. Under coevolution, virus adaptation towards more infectious variants appear considerably faster than when the virus evolves against a static policy. More contagious strains become dominant in the virus population under coevolution. The coevolution regime can generate multiple outbreaks waves as the more infectious variants becoming more dominant in the virus population. Seeing more infectious virus variants becoming dominants may signify that our policy measures are effective.

This artificial coevolution system may offer us a laboratory to "debug" our current policy measures, identify the weaknesses of our current strategies, and anticipate the evolution of the virus to plan ahead relevant policies. It highlights how human behaviors can shape the evolution of the virus, and how reciprocally the evolution of the virus shapes the adaptation of public policy measures. To overcome the simplifications of the implementation in this article, several key innovations for evolutionary algorithms may be required, in particular bringing more advanced biological and genetic concepts in current evolutionary algorithms.

We first present in Section 2 the concept of coevolution, both generally in complex systems, and specifically in our study of the evolution viruses and policies. We propose genetic algorithms as a modelling tool for this context. Genetic algorithms are briefly introduced in Section 3. We present our perspective of using genetic algorithms to generate an artificial coevolution of SARS-CoV-2, and present its main concepts and design in Section 4. Then, we propose an example of implementation of a dual genetic algorithm to model this coevolution process in Section 5, describing the model, the operators, the parameters, and some key results. We develop further the implications and perspectives of this work in Section 6. Section 7 presents data and code availability, and Section 8 concludes.

2 Coevolution of virus traits and policy actions

Coevolution in complex systems

Co-evolution opens a promising and new way to model such ecosystems. Investors in the stock market evolve financial strategies to obtain higher profit, and this evolution can be captured by a GA model. But they are evolving in an environment, that notably includes financial regulations set by policy makers. Not only these regulations are evolving as policy makers strive to identify the best policy to stabilise the market and avoid large crashes: the evolution of regulation and financial strategies is a co-evolution of two species. Policy makers attempt to discourage new loopholes exploited by investors that set a threat on the real economy; investors adapt to the new regulations seeking for other ways to extract profit, finding new niches that trigger new adaptations of regulations. By capturing this interplay, a GA approach could act as a *debugging tool* for financial regulations, a *stress-test program* that invents novel ways to challenge our organisations.

Most sports competitions see such interplay between rules and strategies. The 2008 Olympic Games saw controversy over new swimming suits with novel materials that allowed unprecedented speed and records, leading to their ban causing a change in the innovation strategies of manufacturers. This new direction may spark some day a similar story, calling for new regulation, sparking a different evolution trajectory. Formula 1 constructors actively seek grey-area zones in the regulation hoping for marginal performance gains. One team creatively bypassed the action of a regulatory sensor to increase its engine power, pushing the regulations to add a second sensor and regulate the use of engine modes, impacting all teams' performance. Another racing team exploited unclear rules on purchases and copying of other cars' parts to, leading to a change in the regulations that impacts the evolution of other teams development programs, and that may as well create further unclear rules to be abused in the future. Another

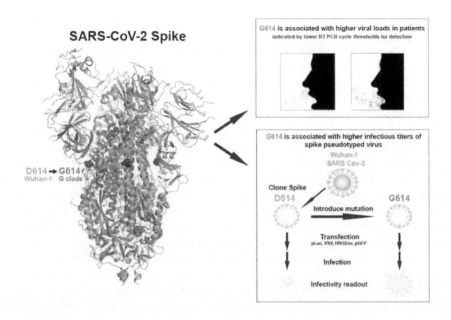

Figure 2: Illustration of the mutation leading to the variant D614G; (Los Alamos National Laboratory, 2020)

instance of coevolution in complex systems, of public high interest, is the co-evolution of viruses and population behaviors or policy measures.

The coevolution of SARS-CoV-2 and policy measures

The emergence of viruses' mutations is a complex topic, both in the mechanisms involved at the virus genome level, but also on what causes some particular mutations to appear, or to be rewarded. That is, the fitness (dis)advantage of the new trait encoded by a mutation, in its environment. We can see the struggle between SARS-CoV-2 mutations illustrated in Figure 2, and human behaviors and policy measures, as an arms race, a coevolution. Humans adopt new restrictions, wear face coverings, adopt social distancing measures, develop testing methods, to reduce the fatalities and infections due to the virus. Facing this pressure, the virus' mutations unconsciously strive to change its genome in order to improve its chances of survival. As some mutations allow the virus to get new, beneficial traits, possibly higher transmissibility (Priya and Shanker, 2021), resistance to antibodies (Callaway, 2020) or causing anomalies in PCR tests (WHO, 2020a), human behaviors may adapt, continuing the arms race. This evolutionary change in traits of individuals in one population, in response to a change of trait in a second population, followed by a reciprocal response, is a phenomenon known as coevolution (Janzen, 1980). Viruses are walking on the fitness landscape (Wright, 1931), a physical representation of the relationship between traits and fitness, and humans change by their behavior this fitness landscape. If by example all humans were hypothetically wearing perfectly hermetic face coverings, airborne transmission methods would fail, causing the virus either to go extinct, or to find other means of transmission.

The continuous interplay between individual genomes or characteristics, and their environment, is an endless source of novelty and niches for adaptation. Individuals are influenced by their environment, and the environment itself is influenced by individual. This dynamic is difficult to model, especially in our context of virus and policies coevolution. The space of possible actions or policy measures is at least very large. Humans can adopt a large diversity of measures, with many levels of stringency or public support. Likewise, the large size of the space of possible genomes for viruses, and the diversity of phenotypes, i.e. observable characteristics, that they can exhibit, challenge our modelling attempts. Coevolution can give birth to novel traits that did not exist before, in a quasi open-ended process. Random or enumerative search methods struggle to evaluate such a large number of possible combinations. We propose here an alternative framework to simulate this coevolution phenomenon in spite of the complexity of the task. Modelling coevolutionary dynamics has seen a large variety of approaches: stochastic processes mathematical modelling (Dieckmann and Law, 1996, Hui et al., 2018), network science (Guimaraes et al., 2017), dynamical systems (Caldarelli et al., 1998), and more biological or genetic methods (Gilman et al., 2012). Evolutionary algorithms (EAs, used for coevolution with Rosin and Belew, 1997), in particular Genetic algorithms (GAs), offer one promising approach at this end. Let us first introduce them briefly, before outlining the properties that makes them relevant for this task.

3 Evolutionary and Genetic Algorithms

A genetic algorithm (GA) is a member of the family of *evolutionary algorithms* (EAs), that are computational search methods inspired from natural selection (Holland, 1992). They simulate Darwinian evolution on individual entities, gathered in a *population*. Genetic algorithms represent these entities with a *genome*, i.e. a collection of genes, often represented as a bit string, that determines the entity *phenotype*, i.e. observable characteristics. The entities undergo selection based of fitness, reproduction of fittest entities, mutations of the genome, that affect their traits (Mirjalili, 2019). Iterating this simplified evolution process, the characteristics of the entities may change, improving the fitness of the population.

As a population-based search method, GAs are efficient in the exploration of *search spaces*, i.e. space of possible solutions, that can be very large

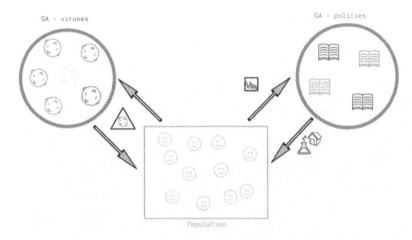

Figure 3: The coevolution model with two genetic algorithms.

(Axelrod, 1987), or rugged (Wiransky, 2020). That is, that admit several extrema, or very irregular structure. They quickly identify regions of the search space that are associated with higher fitness, showing satisfying optimisation capacities (Bhandari et al., 1996). They can also be used to model evolutionary systems, from economies and financial strategies to biological ecologies. Vie (2020a) reviews in more detail its qualities and perspectives as a search method and a modelling tool.

4   An artificial coevolution of SARS-CoV-2

Provided we can formulate an adequate representation of i) the virus genome and ii) policy measures, and under the assumption that the mappings a) between the virus genome and the virus phenotype and b) between the policy actions and the virus phenotype fitness, can be modelled in a satisfying way, we can represent their coevolution as a dual genetic algorithm with two populations: a population of viruses, and a population of policy measures. Both interact indirectly on a third population: the general human population. Viruses survive by infecting new humans in that population, and policy measures modify -to some extent- the behavior of the human population, as Figure 3 illustrates.

   Why GAs? Genetic algorithms are relevant tools to model this coevolution relationships for several reasons. First, evolutionary algorithms appear relevant to model natural selection contexts, as this is precisely their main focus (Holland 1992), though a significant fraction of the literature has used this method for optimisation. Second, among evolutionary algorithms, the inner genetic-centered approach of GAs give them an adequate baseline to encode more complex genomes and phenotypes. The computational architecture of GAs centered on a genetic representation, subject to evolution operators, appears to be the closest to the biological objects we are here interested in modelling. Third, genetic algorithms are particularly powerful in exploring new regions of large search spaces (Whitley, 1994), that may have non trivial structure (Wiransky, 2020). In our coevolution context, we are interested to see what new features may emerge from both viruses and policy responses. GAs, that can generate this novelty, thus constitute a relevant option. Fourth, coevolution has already been modelled using GAs for optimisation (Potter and De Jong, 1994, Vie, 2020b), giving solid foundations for further work in the area, and existing tools to understand the complex dynamics of the artificial SARS-CoV-2 coevolution.

How could this artificial coevolution be implemented? Starting from initial conditions constituted by i) a population distribution of SARS-CoV-2 variants with identified genome sequences and traits and ii) a distribution of the current policy measures, we can simulate the evolution of viruses and policy actions, in response one to another.

To define fitness in this world, one could assume that viruses simply aim at surviving, and do not have an objective function defining some metric to maximise; the performance of policy measures could be evaluated by minimising the number of deaths or infections.

The source of novelty in this coevolution system would essentially be mutations for viruses, and both mutations and recombination for policy measures. While viruses infect new hosts, and don't reproduce between themselves, it is reasonable to consider that national policy makers are exchanging, taking note of what happened in other countries, and changing their own actions in response to positive effects.

From this starting condition, and under these evolution criteria and mechanisms, a large number of runs of the system could be simulated. By observing the behavior of the artificial viruses and policies, and the outbreak dynamics in the artificial human population, some insights could emerge. We could discover some regularities, such as seeing whether and when viruses evolve towards greater transmsissibility, but also observe the changes in the genome, providing useful indications on where to experimentally look at during physical genome sequencing.

This artificial coevolution system may offer us a laboratory to "debug" our current policy measures, identify the weaknesses of our current strategies, and anticipate the evolution of the virus. If a significant portion of the simulations produced viruses that find a way to not be detected by PCR tests, or to evolve a resistance to our current vaccines, policy makers could be advised in advance of this possibility, and work ahead to prevent this issue from happening.

At times where policy makers faced significant uncertainty on the impact of their measures, a difficulty exacerbated by the rather long incubation time of SARS-CoV-2 (Lei et al., 2020), this artificial coevolution system can provide them with a complementary way to assess the impact of prospective policy measures, with an emphasis given on the evolution of the virus. In other words, such simulation possibilities may give the policy maker not only an estimate of the impact of the measures over infection rates and death rates, but also the possibility to consider the consequences of such measures over the future possible traits of the virus.

5   An example of implementation

In this section, we present an implementation example of a coevolution model with dual genetic algorithms. We highlight the building blocks of the model, the parameter configuration, and the key results.

Model

Genetic representation   Individual viruses' genomes in the model are represented as a binary string whose length is the *virus size*. Viruses are initialised with a genome composed exclusively of zeros: this assumes that at the start, viruses are an original form of the disease with no mutations. Each element of this genome represents activation (if equal to 1) or non-activation (if equal to 0) of specific mutated genes. Each mutated gene has an effect on the virus reproduction rate. These effects are drawn uniformly in the interval [-1,1]. This means that some mutations will be detrimental to the virus reproduction, others will have very small or null effects, and some will favor reproduction. We simplify as such the process and effects of mutations, collapsing all these dimensions onto the virus reproduction rate. The virus population contains a given number at the start, programmed by the parameter *initial virus size*.

Individual policies are represented as a binary string as well, initialised with only zeros. This illustrates a starting point in which government policies start with no measure at all. Each element of the policy genome is a policy that can be activated (for a value of the corresponding genome location to 1). Again, we restrict our attention on the virus reproduction rate, and ignore all other dimensions. Each measure will have an effect over the virus reproduction rate, illustrating the efficiency of different measures to prevent the spread of the disease. The effects of these measures are calibrated from the values obtained by Haug et al. (2020) in their influential analysis of the impact of non pharmaceutical interventions. Our model captures the uncertainty on the effects of these policies by setting the effect to be drawn uniformly from the 95% confidence intervals identified by Haug et al. (2020), illustrated in Figure 4. This draw is done once at the beginning of the run. The number of policies considered is parametrised with the *policy population size* parameter, and will remain constant during the run. Policies can include up to 46 measures, corresponding to the measures studied by the above reference.

Infection process   We adopt in this illustration a very simplified model of infection. Each individual virus in the population is characterised by a reproduction rate that incorporates two elements. First, a *"base" reproduction rate*, corresponding to the reproduction rate of the original SARS-CoV-2. Second, this base rate is added to the sum of the effects of mutations activated by this particular individual virus' genome. In the infection step, each virus will infect as many hosts as its *effective reproduction rate*. This effective reproduction rate is equal to the virus reproduction rate, minus the average reduction in reproduction rate in the policy population.

For each new infection, random mutations will happen with a given probability: the *virus mutation rate*. Each element of the virus genome can mutate independently. Higher mutation rates will lead the virus to mutate more frequently during infections. The mutation operator will transform the given element of the genome to a 1 if it is characterised by the value 0, and inversely. As a result, and as the pandemic grows or diminishes, the size of the population of viruses handled by the genetic algorithm will vary, and some diversity may appear within this population.

Fitness   In this model, we reduce the decision makers' problem to a minimisation of the reproduction rate of the virus, which essentially encompass objectives of reduction of deaths. Each individual policy is characterised by a total reduction in the reproduction rate, equal to the weighted sum of the effects of the activated specific measures. The fitness, or value of each individual policy, will evaluate the weighted effective reproduction rate of three viruses chosen at random in the virus population, in a tournament selection process. The policy reduction in the reproduction rate will be applied, and the net, effective reproduction rate recorded. Policies that obtain lower effective reproduction rates will be more likely to be selected in the creation of the next generation of policies.

Viruses do not mutate with an objective. Hence, we have not included a fitness function for the evolution of viruses. Mutations remain unguided by any objectives. The changes of the population of viruses will be driven by the differential reproduction rates of various strains, as described below.

Policy learning   After the fitness of the policies has been determined, policies will be selected to form the basis of next generation policies using "roulette wheel" cumulative fitness selection. Each policy's selection probability will be equal to the ratio of its adjusted fitness (equal to $\frac{1}{1+r}$ where r is the effective reproduction rate of the policy) to the sum of adjusted fitness scores. This crossover step models a process of communication between successful policies: decision makers observe their peers in other countries, observe the measures they implement and the associated results. Measures that appear efficient abroad tend to be implemented nationally by the means of this imitation step. This crossover step occurs with probability equal to the *policy crossover rate*. After selecting two policies, a random

Parameter	Value
Virus initial population size	10
Virus size	10
Policy population size	100
Base reproduction rate	2.63
Tmax	20
Policy crossover rate	0.5
Policy mutation rate	0.05
Virus mutation rate	0.0001

Table 1: Parameter configuration for the dual genetic algorithm

uniform crossover point will be determined, and the two policies' genomes will be interchanged after this crossover point. The result of this procedure will be two children policies for the next generation. Otherwise, when the crossover operator is not activated with probability 1 - *policy crossover rate*, the children strategies will be exact copies of their parents.

Learning to improve policies will also include a mutation step, modelling small perturbations or explorations. This illustrates for instance a country implementing or removing quarantine restrictions for various reasons. With a *policy mutation probability*, any element can mutate from value 0 to value 1. We outline here one important limitation: we do not allow policies in our model to revert back after some measures have been implemented: we essentially forbid detrimental mutations. Extending our space of possible measures to measures that do not work could be an interesting direction as well. We also do not consider other factors such as economic output or political situation that could act as a pressure towards relaxation of measures. Again, these constraints would be an interesting addition for this model, but we have chosen to present a simple illustration of coevolution.

Evolution run and parameters   The simulation runs for *Tmax* periods. We run our simulations for a base reproduction rate of 2.63 (Mahase, 2020). Note however that simply changing the value of the base reproduction rate, or including uncertainty on its determination, is easily achievable in the source code (see below for availability). Higher base rates will likely make the infection spike faster and higher, while lower base rates may lead to the virus extinction in some cases, or reductions of the outbreak peaks. In the model, we consider the time periods to be indexed as weeks, assuming that each virus is transmitted every seven days.

A situation of coevolution defines a run in which both the viruses and the policy can evolve: that is, their mutation rates and the policy crossover rate are strictly positive. When the virus mutation rate is null, but the policy mutation rate and policy crossover rates are positive, we model a situation in which only the policy is evolving, against a static virus. When the virus mutation rate is positive, and the policy mutation rate and crossover rates are null, we are illustrating a situation in which the virus evolves, and policies remain indifferent and void. All other parameters remain unchanged.

Before turning to the simulation results, we make a note on the impact of the parameters over the results, and the outbreak dynamics that are generated. A major challenge in this example implementation was to avoid too large epidemics: as each virus is simulated individually, handling hundred of millions of viruses can incur a significant computational cost. The development of the simulation allowed us to be able to simulate in reasonable time (seconds) up to ten billion individual viruses. Higher virus mutation rates, or higher initial virus sizes, or less effective policies, can lead to exponential growth of the virus population size. Alternatively, if policies are very efficient (high mutation rates and crossover rates), and if the virus does not mutate frequently enough, the model may manage to make the virus go extinct. We must acknowledge that simulation results can be sensitive to small variations of the parameters. The configuration showed in Table 1 allows to keep computation doable for the 20 time periods considered. Outside extreme situations (complete virus takeover or virus extinction), the main insights presented below hold.

Results

We now run the evolution of viruses and policies in these three situations above, to identify specific features of the coevolution regime. The Figure panel 5 presents the main results. Their observation allows us to formulate a few "stylized facts" of the coevolution of viruses and policies.

1. **Under coevolution, virus adaptation towards more infectious variants is considerably faster than when the virus evolves against a static policy.** In Figure 5a, we can observe that the average reproduction create in the virus population rises to 3.1 after 20 time periods under coevolution (red curve). When the virus does not evolve (blue), the average reproduction rate naturally stays at the initial value of 2.63. Interestingly, when the virus can evolve, but when the policy does not (green curve), the average reproduction rate tends to increase slightly, but much less than under the coevolution regime. Having the virus face a more severe struggle for its survival makes its evolution more efficient.

2. **More contagious strains become dominant in the virus population under coevolution.** Figure 5d shows the frequency of viruses in the virus population containing the mutation gene granting the highest increase in reproduction rate. This fraction rises to 0.35 in the

Fig. 1: Change in R_t (ΔR_t) for 46 NPIs at L2, as quantified by CC analysis, LASSO and TF regression.

From: Ranking the effectiveness of worldwide COVID-19 government interventions

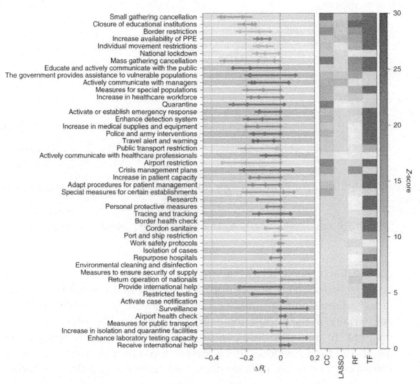

Figure 4: Effects of Covid 19 government interventions (From Haug et al., 2020). With permission from Nature Human Behavior - Reproduction License 4994130245697 (Jan. 22 2021)

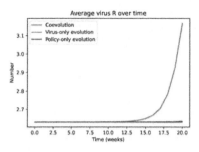

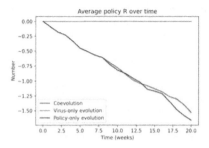

(a) Average reproduction rate of the population of viruses over time.

(b) Average impact in reproduction rate of policies over time.

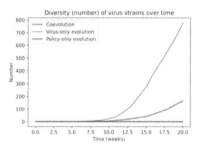

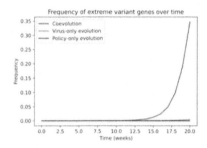

(c) Number of different virus strains over time.

(d) Frequency of extreme variant genes over time.

Figure 5: Key results from the coevolution dual genetic algorithm.

coevolution case, while this share is considerably lower under virus-only evolution. This point supports the idea that coevolution makes virus' adaptation much more efficient. Indeed, the number of different variants in the population exposed by Figure 5c shows interesting insights. In the virus-only evolution, up to 800 variants appear during the 20 time periods. This is due to the outbreak dynamic: in the virus-only evolution, policies do not do anything and do not change, hence the virus is free to spread everywhere. As its population size grows, more mutations happen, and more variants emerge. Under coevolution, only up to 200 variants emerge, but the frequency of the strongest mutations shows that virus evolution is made much more efficient by the challenge proposed by learning policies.

3. **The coevolution regime can generate multiple outbreaks waves as the more infectious variants becoming more dominant in the virus population.** While currently in European countries, a so-called third wave seem to have occurred coincidentally to the VOC 202012/01 (the "UK variant") becoming dominant, this pattern occurred as well dur-

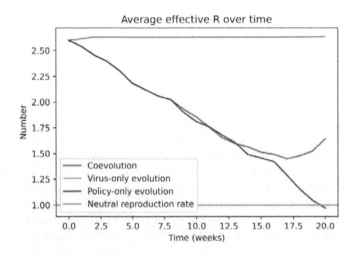

Figure 6: Average effective reproduction rate over time.

ing our evolution run. Figure 5b shows that policies evolve to be more efficient over time, leading the average effective reproduction rate of the virus to go below 1, in a path to extinction. Under the coevolution regime, the more efficient adaptation of the virus allows instead the effective reproduction rate to increase again. Several multiple waves seem empirically to stem from relaxing measures, a behavior that our model does not include. However, the same pattern and insight would hold. In this simulation of coevolution, multiples waves of infection can occur because of increasing viruses' reproduction rates, or relation of policy measures.

4. **Seeing more infectious virus variants becoming dominants may signify that our policy measures are effective**. These sets of figures show that when policies are not evolving and not effective, more infectious variants take a much longer time to become dominant in the population. Only when policies evolve and actively undermine the virus reproduction, weaker forms progressively disappear, to be replaced by stronger virus variants. Several countries today see numerous variants quickly increase in the share of new infections. While this dynamic constitutes a key challenge and difficulty, it can be seen as the sign that the current measures are putting stress on the virus: they are efficient in pushing weaker forms to reduction and eventually extinction. Only by continuously adapting, and adapting faster than the virus strains, can policies and human behaviors push all variants to final extinction. Our future work with this model will strive to include vaccines as a policy measures, allow viruses to obtain a

vaccine-resistant trait by mutations, and observe how the evolution of policies shapes the emergence of vaccine-resistant strains of SARS-CoV-2.

6   Implications and perspectives

This perspective for the artificial coevolution first faces the challenges inherent to the use of GAs, that were recently reviewed by Vie (2020a). Their computational cost increases significantly with the size of the populations they consider. If we wanted to simulate very large population of viruses, knowing that the evolution of SARS-CoV-2 is a hugely parallel process occurring over millions of hosts simultaneously, the computational cost of the simulation would be significant. In addition, small differences in parameter configuration of GAs, including population size, mutation rates, selection intensity, is difficult in GAs, as different sets of parameters may yield different results, and impact the algorithm performance, or convergence properties (Grefenstette, 1986). Last but not least, the genetic representation needs careful design to cover the diversity of possible solutions in a realistic manner, without creating unintended loopholes that could be exploited by the algorithm (Juzonis et al., 2012) and bias the results. Several recent works shed new light on these challenges, and provide new means to mitigate their effects. The computational cost of GAs fades before their great scaling with parallelism (Mitchell, 1988), and the computing power of GPUs (Cheng and Gen, 2019) or Cloud computing hardware. New methods have been introduced in parameter configuration (Hansen, 2016; Huang et al., 2019; Case and Lehre, 2020). A large diversity of genetic representations exist in GAs, and some further inspiration from key biological concepts can open the way to representations allowing these algorithms to evolve more complex artificial organisms (Miikkulainen and Forrest, 2021).

Specifically in the perspective of the artificial coevolution laboratories discussed here, a key challenge remains in establishing a proper algorithmic representation of the SARS-CoV-2 genome, and the mapping between this genome and the virus traits. By proper, we mean that this representation might not need to be comprehensive or perfectly exact, but should not oversimplify the object being studied, or neglect important determinants of traits. The work perspective described here faces important limitations, and as these algorithms could be used for essential matters of public health, the biases they may contain require careful consideration. These programs cannot simulate at perfection natural selection or comprehensive genetics, simply because we do not fully understand them yet.

Attempting to model the coevolution of viruses with more realistic simulations than the example provided here is certainly a challenging endeavor. It however entails significant benefits and opportunities. The recent mutations of SARS-CoV-2 have raised public awareness about this critical issue for public health, and make attempts to address this issue with a matter of public interest, with immense benefits when we consider the cost faced by the general public due to variants-caused restrictions. This challenge constitutes as well an opportunity for evolutionary algorithms to grow. If we can make these computer programs that simulate natural selection capable of representing and simulating the evolution of viruses, which are organisms considerably more complex that what EAs are currently handling, these improved EAs in the future could lead to breakthroughs in bioinformatics, optimisation, artificial life and AI.

How could such algorithms evolve organisms with that level of complexity? Modifications of GAs that move from the simple bit string representation to more complex genomes, can start this transformation. Key phenomena in genetics and biology such as *pleiotropy* -where one gene impacts several traits-,*polygeny* -one trait is impacted by several genes-, the evolution of *evolvability*, realistic mutations, are yet to be included in these algorithms, and their addition carries significant benefits and new opportunities. These "structural" genetic algorithms that place such emphasis on the genome structure, may make us able to evolve much more complex, adaptive artificial entities to study viruses evolution as illustrated here, but also to create advanced forms of artificial life, or foster progress in generative artificial intelligence. The challenge of modelling SARS-CoV-2 coevolution with genetic methods can inspire such decisive innovations.

7 Data and code availability

The main simulation code of the GA proof of concept is freely available at https://github.com/aymericvie/Covid19_coevolution. Model parameters such as the efficiency of different non pharmaceutical interventions, or the basic reproduction rate of SARS-CoV-2, as well as mutation rates, or learning rates for policies, can be easily changed in the code. The code is designed to work on Google Colab, and the script is self sufficient to run.

8 Conclusion

In this article, we propose coevolution with genetic algorithms (GAs) as a credible approach to model this relationship, highlighting its implications, potential and challenges. We provide a proof of concept-implementation of this coevolution dual-GA. Because of their qualities of exploration of large spaces of possible solutions, capacity to generate novelty, and natural genetic focus, GAs are relevant for this issue. We present a dual GA model in which both viruses aiming for survival and policy measures aiming at minimising infection rates in the population, competitively evolve. Under coevolution, virus adaptation towards more infectious variants appear considerably faster than when the virus evolves against a static policy. More contagious strains become dominant in the virus population under coevolution. The coevolution regime can generate multiple outbreaks waves as the more infectious variants becoming more dominant in the virus population. Seeing more infectious virus variants becoming dominants may signify that our policy measures are effective. This artificial coevolution system may offer us a laboratory to "debug" our current policy measures, identify the weaknesses of our current strategies, and anticipate the evolution of the virus to plan ahead relevant policies. It also constitutes a decisive opportunity to develop new genetic algorithms capable of simulating much more complex objects. We highlight some structural innovations for GAs for that virus evolution context that may carry promising developments in evolutionary computation, artificial life and AI.

Bibliography

[1] Axelrod, R. (1987). The evolution of strategies in the iterated prisoner's dilemma. Genetic algorithms and simulated annealing, 32-41.

[2] Benvenuto, Domenico and Giovanetti, Marta and Ciccozzi, Alessandra and Spoto, Silvia and Angeletti, Silvia and Ciccozzi, Massimo (2020). The 2019-new coronavirus epidemic: evidence for virus evolution. Journal of medical virology, 92(4), 455–459

[3] Bhandari, D., Murthy, C. A., & Pal, S. K. (1996). Genetic algorithm with elitist model and its convergence. International journal of pattern recognition and artificial intelligence, 10(06), 731-747.

[4] Callaway, E. (2020). Making sense of coronavirus mutations. Nature, 174-177.

[5] Caldarelli, G., Higgs, P. G., & McKane, A. J. (1998). Modelling coevolution in multispecies communities. Journal of theoretical biology, 193(2), 345-358.

[6] Case, B., & Lehre, P. K. (2020). Self-Adaptation in Nonelitist Evolutionary Algorithms on Discrete Problems With Unknown Structure. IEEE Transactions on Evolutionary Computation, 24(4), 650-663.

[7] Cheng, J. R., & Gen, M. (2019). Accelerating genetic algorithms with GPU computing: A selective overview. Computers & Industrial Engineering, 128, 514-525.

[8] Dieckmann, U., & Law, R. (1996). The dynamical theory of coevolution: a derivation from stochastic ecological processes. Journal of mathematical biology, 34(5), 579-612.

[9] Domingo, Esteban and Escarmís, Cristina and Sevilla, Noemi and Moya, Andres and Elena, Santiago F and Quer, Josep and Novella, Isabel S and Holland, John J (1996). Basic concepts in RNA virus evolution. The FASEB Journal, 10(8), 859–864.

[10] . Duffy, Siobain (2018) Why are RNA virus mutation rates so damn high? PLoS biology, 16 (8), Public Library of Science San Francisco, CA USA

[11] Gilman, R. T., Nuismer, S. L., & Jhwueng, D. C. (2012). Coevolution in multi-dimensional trait space favours escape from parasites and pathogens. Nature, 483(7389), 328-330.

[12] Guimaraes, P. R., Pires, M. M., Jordano, P., Bascompte, J., & Thompson, J. N. (2017). Indirect effects drive coevolution in mutualistic networks. Nature, 550(7677), 511-514.

[13] Grefenstette, J. J. (1986). Optimization of control parameters for genetic algorithms. IEEE Transactions on systems, man, and cybernetics, 16(1), 122-128.

[14] Hansen, N. (2016). The CMA evolution strategy: A tutorial. arXiv preprint arXiv:1604.00772.

[15] Holland, J. H. (1992). Genetic algorithms. Scientific american, 267(1), 66-73.

[16] Holland, J. H. (1992). Adaptation in natural and artificial systems: an introductory analysis with applications to biology, control, and artificial intelligence. MIT press.

[17] Huang, C., Li, Y., & Yao, X. (2019). A survey of automatic parameter tuning methods for metaheuristics. IEEE transactions on evolutionary computation, 24(2), 201-216.

[18] Hui, C., Minoarivelo, H. O., & Landi, P. (2018). Modelling coevolution in ecological networks with adaptive dynamics. Mathematical Methods in the Applied Sciences, 41(18), 8407-8422.

[19] Iacobucci, G. (2021). Covid-19: New UK variant may be linked to increased death rate, early data indicate. bmj, 372, n230.

[20] Janzen, D. H. (1980). When is it coevolution?

[21] Juzonis, V., Goranin, N., Cenys, A., & Olifer, D. (2012). Specialized genetic algorithm based simulation tool designed for malware evolution forecasting. Annales Universitatis Mariae Curie-Sklodowska, sectio AI–Informatica, 12(4), 23-37.

[22] Kai Kupferschmidt (2020). Mutant coronavirus in the United Kingdom sets off alarms, but its importance remains unclear. https://www.sciencemag.org/news/2020/12/mutant-coronavirus-united-kingdom-sets-alarms-its-importance-remains-unclear. Accessed: 31 December 2020.

[23] Korber, B., Fischer, W. M., Gnanakaran, S., Yoon, H., Theiler, J., Abfalterer, W., ... & Montefiori, D. C. (2020). Tracking changes in SARS-CoV-2 Spike: evidence that D614G increases infectivity of the COVID-19 virus. Cell, 182(4), 812-827.

[24] Los Alamos National Laboratory, Newer variant of COVID-19–causing virus dominates global infections (2020). https://lanl.gov/discover/news-release-archive/2020/July/0702-newer-variant-covid-dominates-infections.php. Accessed: 19 February 2021.

[25] Lei, S., Jiang, F., Su, W., Chen, C., Chen, J., Mei, W., ... & Xia, Z. (2020). Clinical characteristics and outcomes of patients undergoing surgeries during the incubation period of COVID-19 infection. EClinicalMedicine, 21, 100331.

[26] Lima, C. (2020) Information about the new coronavirus disease (COVID-19). Radiologia Brasileira, 53 (2), V–VI.

[27] Mahase Elisabeth. Covid-19: What is the R number? BMJ 2020; 369 :m1891

[28] Mahase, E. (2021). Covid-19: Novavax vaccine efficacy is 86% against UK variant and 60% against South African variant.

[29] Matyásek, Roman and Kovarík, Ales (2020). Mutation patterns of human SARS-CoV-2 and bat RaTG13 coronavirus genomes are strongly biased towards C> U transitions, indicating rapid evolution in their hosts. Genes, 11(7), 761.

[30] Miikkulainen, R., Forrest, S. A biological perspective on evolutionary computation. Nat Mach Intell 3, 9–15 (2021). https://doi.org/10.1038/s42256-020-00278-8

[31] Mirjalili, S. (2019). Genetic algorithm. In Evolutionary algorithms and neural networks (pp. 43-55). Springer, Cham.

[32] Mitchell, M. (1998). An introduction to genetic algorithms. MIT press.

[33] Haug, N., Geyrhofer, L., Londei, A., Dervic, E., Desvars-Larrive, A., Loreto, V., ... & Klimek, P. (2020). Ranking the effectiveness of worldwide COVID-19 government interventions. Nature human behaviour, 4(12), 1303-1312.

[34] Phan, Tung (2020). Genetic diversity and evolution of SARS-CoV-2. Infection, genetics and evolution, 81.

[35] Potter, M. A., & De Jong, K. A. (1994). A cooperative coevolutionary approach to function optimization. In International Conference on Parallel Problem Solving from Nature (pp. 249-257). Springer, Berlin, Heidelberg.

[36] Priya, P., & Shanker, A. (2021). Coevolutionary forces shaping the fitness of SARS-CoV-2 spike glycoprotein against human receptor ACE2. Infection, Genetics and Evolution, 87, 104646.

[37] Rosin, C. D., & Belew, R. K. (1997). New methods for competitive coevolution. Evolutionary computation, 5(1), 1-29.

[38] Vie, A. (2020a). Qualities, challenges and future of genetic algorithms: a literature review. arXiv preprint arXiv:2011.05277.

[39] Vie, A. (2020b). Genetic algorithm approach to asymmetrical blotto games with heterogeneous valuations. SSRN.

[40] Wallace, D. J., & Ackland, G. J. (2021). Abrupt increase in the UK coronavirus death-case ratio in December 2020. medRxiv.

[41] Whitley, D. (1994). A genetic algorithm tutorial. Statistics and computing, 4(2), 65-85.

[42] WHO (2020a). SARS-CoV-2 Variants. `https://www.who.int/csr/don/` `31-december-2020-sars-cov2-variants/en/`. Accessed: 31 December 2020.

[43] WHO (2020b). SARS-CoV-2 mink-associated variant strain – Denmark. `https://www.who.int/csr/don/` `06-november-2020-mink-associated-sars-cov2-denmark/en/`. Accessed: 31 December 2020.

[44] Wirsansky, E. (2020). Hands-On Genetic Algorithms with Python. Packt Publishing.

[45] Wright, S. (1931). Evolution in Mendelian populations. Genetics, 16(2), 97.

Logistic Growth Model in Epidemics

M. Bahrami* and R. H. Rimmer

Wolfram Research, 100 Trade Center Dr, Champaign, IL, 61820, USA.
*Corresponding author: mohammadb@wolfram.com

A logistic function for population growth can be used for modelling epidemics controlled only by effective measures such as quarantine. The model counts numbers of cases or deaths as the measure of viral growth. Effective quarantine measures separate the healthy population from the exposed population, limiting spread of the virus. To be effective, the quarantine must be accomplished more quickly than the virus can spread through the population. While the logistic model appeared to fit the early COVID-19 data from a few countries, it failed as the virus spread to other countries. The most useful application of the model may be to test the effectiveness of quarantine methods. Keywords: Logistic function, COVID-19, population growth, nonlinear fitting.

1 Introduction

The logistic equation is very simple description of a population growth proposed by Pierre François Verhulst 180 years ago [1]. It can also describe an epidemic outbreak, expressed as cumulative number of cases or deaths, when the primary method of control is quarantine as in the case of a novel viral infection. Cumulative cases or deaths can be treated as crude measures of the virus population growth, and the quarantine method is to stop the spread of the virus by removing its substrate which is the human population at risk, which is initially assumed to be the entire uninfected population [2–8].

There are many sophisticated epidemiological models which separate a population at risk into compartments of susceptible, exposed, infectious, recovered, deceased and so forth (see Ref. [9] and references therein). Each compartment is modeled by a differential equation with its own rate parameters and potential initial conditions. Early in an epidemic with a novel infectious agent, most of these parameters are not known, making predictions about the epidemic difficult. The logistic model can be represented by a single differential equation with two parameters and one initial condition. Ideally this simplicity should allow early predictions about the rate of spread and limits to the epidemic by effective quarantine methods.

In this paper, we reviewed the underlying mathematical details of the logistic function. We then compared its predictions with the actual COVID19 data across different countries. We finally discussed the applicability of this model, and how it can be used to evaluate the effectiveness of quarantine methods.

2   Methods

The logistic growth model is defined by a differential equation as follows:

$$\frac{df(t)}{dt} = kf(t)\left(1 - \frac{f(t)}{L}\right),$$ (1)

where k is a continuous growth rate and L is the limit to growth such that $f(t)$ cannot exceed L. The parameter k is also called as the Malthusian parameter (i.e., the rate of maximum population growth) and L as the so-called carrying capacity (i.e., the maximum sustainable population). The parameter k determines the rate of infection spread and how long the epidemic will last. The parameter, L, is the limit of the number of infected people or the subset of people who become infected or die. For more info, see Ref. [1,2] and references therein.

At early stages of an outbreak when $f(t)$ is small relative to L, the term, $1 - f(t)/L$, is near 1. Then the initial growth is nearly proportional to $f(t)$ with exponential growth rate, k. When $f(t)$ reaches L, the growth rate is zero and growth stops. For positive L and k in Eq.1, the growth rate is always positive. Therefore, Eq.1 has an explicit solution:

$$f(t) = \frac{L}{1 + e^{-k(t-t_0)}}.$$ (2)

where the t_0 parameter is the time of the peak of the epidemic on an arbitrary time scale[1].

Additionally, one can replace the logistic differential equation by a general iterative form which is usually called as the logistic map:

$$\begin{aligned}
\text{continuous form}: \quad x' &= rx(1 - x), \quad \text{with} \ x = f(t)/L, \\
\text{discrete form}: \quad x_{n+1} &= rx_n(1 - x_n),
\end{aligned}$$ (3)

This quadratic map, as a quadratic recurrence equation, can show a very complicated behavior. Note the parameter r is sometimes called the biotic potential [12]. As we will discuss later, the value of parameter r for COVID-19 data is less than 1, which is in the regime where Eq.(3) does not show chaotic features.

The logistic function also resembles the growth rate of autocatalytic chemical reactions, where the maximum amount of $f(t)$ cannot go beyond

---

[1]Note t_0 is not initial time. It depends on the initial value of f, i.e. $t_0 = \ln(L/f(0) - 1)/k$.

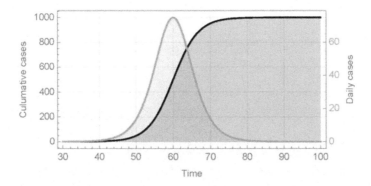

Figure 1: The logistic function (black) and its first derivative (orange) for $k = 0.3$, $L = 1000$, and $t_0 = 60$.

the initial concentrations [10, 11]. In its simplest form, the corresponding reaction and the rate law is given by:

$$A + B \longrightarrow 2B, \qquad \frac{d[A]}{dt} = k[A][B], \tag{4}$$

with k the rate constant and $[X]$ the concentration of chemical X. Given the stoichiometry of reaction, we have $[A]_0 + [B]_0 = [A]_t + [B]_t$, with $[X]_t$ the concentration of chemical X at time t. Assuming $[A]_0 + [B]_0 = L$, one finds:

$$[B]_t = \frac{L}{1 + e^{-r(t-t_0)}}, \tag{5}$$

with $t_0 = \ln([A]_0/[B]_0)/r$ and $r = kL$. The above equation resembles the solution of the logistic differential equation as in Eq.(2).

As one can see in Fig. 1, the logistic function predicts one symmetric wave (i.e., one surge in the number of infected cases). What we have experienced so far across the world has been not only second or more waves, but almost always asymmetric and heavy-tail waves (see Fig. 2). One may argue that the logistic growth model has failed badly. However, the great convenience of the model is that with only three parameters, the function can be described. Once quarantine measures in the course of the epidemic are fully implemented and the case rates have stabilized, the parameters should be relatively constant as the epidemic progresses to its conclusion.

At the start of an epidemic, there is a large cohort of susceptible population with nothing to limit growth. It is reasonable to assume that growth of the number of cases would be exponential, i.e. the instantaneous rate of growth will be proportional to the current number infected:

$$f'(t) = kf(t), \tag{6}$$

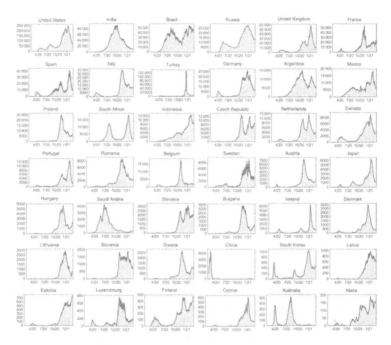

Figure 2: Daily COVID19-confirmed cases for G20-member countries vs time. The date tick labels are in two-letter month/two-letter year format. For the COVID19 data, we used the curated data freely available in Ref. [13].

which shows an exponential growth, with $f(t)$ being the cumulative case function of time. If we compare it with Eq.(4), the early stages of an epidemic resembles a pseudo first-order reaction: it is not strictly a first-order process (i.e., exponential growth), but it appears to be so because the population of susceptible[2] is by far much larger than the infected (i.e. $[A] \gg [B]$ in Eq.(4)).

The logistic model makes no assumptions about immunity, developing resistance, deaths, vaccination, or recovery, or anything reducing the susceptible population. The model can only be expected to work for an epidemic that is quickly controlled by quarantine. To modify the exponential growth to account for quarantine, we multiply the exponential growth by a probability of transmission of the virus as a function of time, $S(t)$, which will range between 0 and 1 and is essentially a survival function for the virus at time, t:

$$f'(t) = kf(t)S(t). \tag{7}$$

If quarantine methods are effective, persons known to be infected or exposed to the infection are isolated from the general population and from each other to prevent further spread both in the quarantine group and the general population. For the quarantine method to succeed the number of quarantined persons must grow faster than the actual infected population so that eventually the infection is limited to the quarantined individuals and can spread no further. An approximate function for the probability of transmission at time, t, can then be modeled with the function:

$$S(t) = 1 - \frac{f(t)}{L}. \tag{8}$$

Note the number L will not be known until the dynamics of the early epidemic are known. Therefore, early in the epidemic, the probability of transmission will be near 1, but it will rapidly approach zero as $f(t)$ approaches L, because $f(t)$ is growing exponentially. Multiplying the equations, we get the logistic differential equation as in Eq.(1).

3   Results and Discussions

In this section, we focused on only one surge of COVID19 (i.e., one wave) across some countries to explore the predicative power of the logistic model. For each case, we considered the daily confirmed-cases, and implemented fitting using `NonlinearModelFit` function in Mathematica.

South Korea: a good fit

Fig. 4(a) shows the accumulated case data for South Korea. The very early data show a very small outbreak confined to 31 cases. The major outbreak

---

[2]Note we mean the *actual* susceptible population, L, which is not necessarily the overall population.

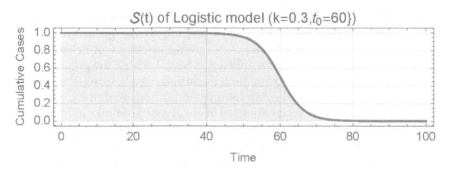

Figure 3: The probability of transmission of the virus as a function of time in the Logistic model, $S(t)$.

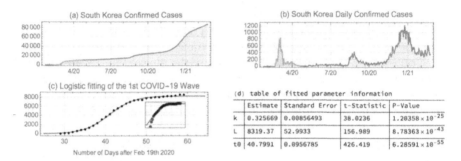

Figure 4: South Korea COVID19 data: (a) the cumulative cases, (b) the daily ones, (c) the logistic function fitting with the first wave, and (d) the corresponding fitted parameters. The inset in (c) is the same plot in log-scale. Note the 1st wave is highlighted in (a,b) The South Korea data is obtained from Ref. [13].

had an epicenter with a small group that had visited Wuhan Province. The data from this group showed up after 19 Feb 2020. Since the origin of these cases was confined to a small known group, quarantine measures were relatively easy to enforce and the data followed the logistic model until mid March.

We select the data between 20 Feb and 20 March for analysis. For fitting purposes day 0 of the arbitrary time scale is taken to be 21 January 2020, as the data source begins on the following day. The parameters of the fit for this subset of data are shown in Fig. 4(d). The last two data points were breaking away from the model, showing a linear growth of cases which continued in the subsequent points, indicating the quarantine efforts were no longer fully effective. The logarithmic plot (in the inset of Fig. 4(c)) usually gives an idea if the data are appropriate for the model. Initially the data should be nearly linear on the early points, although it is not uncommon for there to be counting and reporting problems with the very early data points. The curve begins to bend before the inflection point of

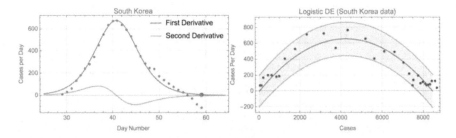

Figure 5: The first and second derivatives of the logistic function (left), and the corresponding fitting with the logistic DE in Eq.(1) (right). The green dots represent points calculated from the data points, using the differential equation in Eq.(1), and the values for k and L determined by the fit. The red dot (left graph) is the location on the first derivative curve at the last day of the data.

the sigmoid curve and the curve quickly thereafter becomes horizontal as L is asymptotically approached.

Fig. 5 shows the first and second derivatives of the logistic function for the same data as in Fig. 4(b). The peak of the epidemic occurs when the second derivative curve crosses zero. The data can also be fit to the model using the logistic differential equation. To do this a method is needed to calculate the first derivative from the data. In the graph the first derivative is calculated using an interpolation function, but first differences are usually satisfactory even if they are offset by a half day. The red dots are the first derivative points calculated at the data point. The fit parameters are similar to the fit to the cumulative data.

Italy case: testing the predictive power of the model

As discussed before, the COVID19 waves across the world are usually highly-asymmetric, therefore the logistic model fails to describe this feature, although the trajectory before the peaking of waves are relatively well-described by the logistic function. This issue is shown in Fig. 6, which illustrates the first surge of COVID19 cases in Italy through Mar-Jul 2020 (123 days). For fitting and estimating parameters, we have only considered the weekly moving-average data for the first 20 days (black points). As one can see, after peaking, the number of cases drops much slower than what the logistic model predicts; in other word, the wave is fat-tailed, which implies the survival function for the virus in the logistic model drops much faster than the actual data. We have explored this issue quantitatively in the following section.

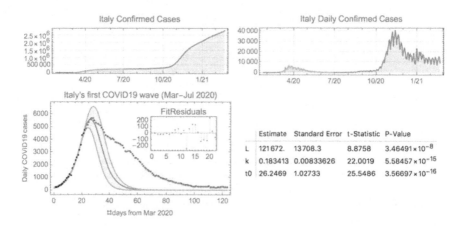

Figure 6: Italy COVID19 data cases. The red dots in lower graph are actual data and the solid ones are the fitted logistic function (orange line is the 90% confidence band). On the right, one can see the table of fitted parameters. The COVID19 data is obtained from Ref. [13].

Stability of parameters in the logistic model

Here, we briefly explored the parameters stability of logistic model for Italy's COVID19 data. The basic idea of the quarantine model is that there is an effective quarantine. If that were true, parameters should start to converge quickly. The average incubation period is only about a week. If all the contacts of every person were isolated quickly enough, the epidemic should end in a few weeks. If quarantine is partial, the epidemic does not ever end until the whole population is infected or immunized. Starting from Feb 2020, as the day number increases, an additional day is added to the sample we used for fitting. Clearly there is not clear parameter stabilization (see Figs. 7 and 8).

As the k parameter decreases, the logistic model becomes more pro-tracted. If k keeps declining a peak will never be reached. In other word, a slight increase in the value of k is much better observation (more effective quarantine) than a decline (look at Fig. 7 after 250 days).

After the inflection point of the first wave, which is about 50 days (see Fig. 7), the parameters L and t_0 reaches some stability, which is completely lost after 250 days with the beginning of the 2nd surge (2nd wave in Fig. 6). As one can see in Fig. 8, the parameter L shows larger volatility compared to t_0, implying very likely a big change in the underlying dynamic of virus spread (e.g., new communities being infected).

Additionally, as the surge in cases starts its exponential increase, one observes a steady decline in k while L increases slowly. As we approach the peaking of the wave, L increases dramatically and then decreases while k remains more or less constant. After wave peaking, there won't be that

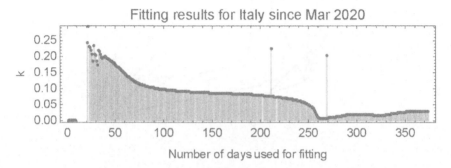

Figure 7: The best fitted value of k vs different number of days used for fitting. The x-axis denotes the number of days after Feb 1st, 2020 that were used for fitting.

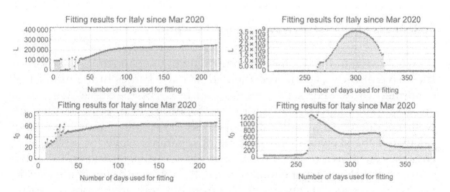

Figure 8: The best fitted values of L and t_0 given different days used for fitting. The x-axis denotes the number of days after Feb 1st, 2020 being used for fitting. Note due to a jump in values, we have plotted the first 220 days separately (left).

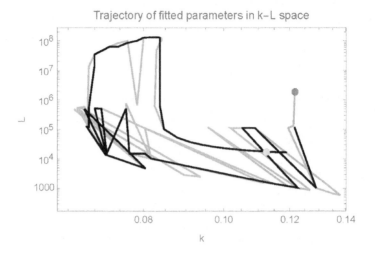

Figure 9: The trajectory of fitted parameters k, L for the first wave of Japan (02/2020-06/2020). The red dot is the fitted values for the first 10 days, and the green one for 120 days, starting from 02/2020. The black solid line in the moving median, for better visualization of the trajectory.

much change in L and one observes a steady increase in the fitted value of k. Fig. 9 shows such a behaviour for the 1st wave of COVID19 cases in Japan.

Tailedness of COVID19 waves across the world

As we showed in Fig. 2, since the beginning of the pandemic, many countries have already experienced the 2nd or more surges in the number of confirmed cases. For some countries such as US, the successive waves are greatly overlapped, while for other countries such as Japan, they are well-separated. To quantify the tailedness of separated waves, we have calculated the skewness and kurtosis of separated wave for a sample of countries as shown in Fig. 10. As one can see, almost always, the wave starts with a sharp exponential increase, then decrease slowly after peaking. So as expected, it is very different from not only predictions of logistic function, but also almost all other epidemiological modelings. Our main goal here is not to forecast using the logistic function, but to understand the underlying dynamics of a pandemic, e.g., the effectiveness of preventive measures such as quarantine, and we believe the logistic function, although extremely simple, can help us in this direction.

One approach to improve the logistic model could be phenomenological modifications (e.g., see Ref. [15] and references therein). For example, one

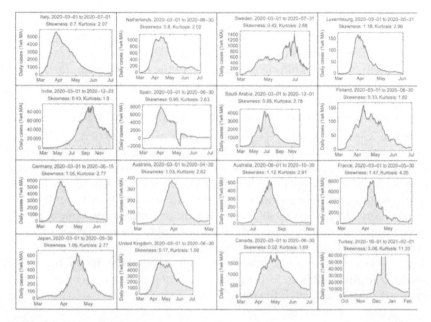

Figure 10: Kurtosis and Skewness of COVID1-9 surge waves (i.e. daily COVID-19 confirmed cases) for some countries with well-separated waves. To calculate these measures, we have used the weekly moving-averaged values.

can generalize the corresponding differential equation as follows:

$$\frac{df}{dt} = k(t)f^p\left(1 - \frac{f}{L(t)}\right), \tag{9}$$

with p a growth scaling parameter. To remain faithful to the analogy between the logistic function and autocatalytic reactions, we think setting $p = 1$ and exploring the time behaviour of k and L is more interesting. The predictions of above equation seem to be more sensitive toward changes in L rather than k, however, the actual behaviour highly depends on the interplay of these two. For example, slight increase in L after the peaking of wave, together with slight decrease in k results in a highly-asymmetric wave, whose fatness depends on the actual values of k and L and how they change in time.

Concluding remarks

The logistic model is extremely simple, using only quarantine to limit the epidemic and making no assumptions about either the virus or its human substrate. It always takes authorities several incubation periods to determine that an epidemic outbreak has occurred, so quarantine methods lag the spread of a virus. During this lag period, the probability of transmission of the virus is almost constant, $S(t) = 1$, and the solution to the logistic differential equation shows exponential growth with the same growth factor, k. The growth factor magnitude is a function of the rate of transmission of the virus and human behavior and susceptibility. The parameter k should not change as the epidemic progresses and quarantine is implemented. With the logistic model, $S(t)$ begins to decline with time only after quarantine is effective. Since $S(t) = 1 - \frac{f(t)}{L}$, it will decline more quickly if k is large, as $f(t)$ will more rapidly approach L. The consequence for the logistic model is that, if k is large, the epidemic will end quickly if and only if the quarantine is effective. A large magnitude for k creates a problem for authorities implementing the quarantine, as they have to trace contacts faster than the virus is spreading. COVID-19 has the insidious property that many people have symptoms no worse than the common cold, so they may spread new outbreaks which will not be discovered until more vulnerable individuals develop serious symptoms. When the serial fitting of the data to the model shows k declining, it is an indication that the quarantine is incomplete, and the model does not match the data.

Acknowledgments.

Authors thank the referees for their comments, helping us to improve this paper. M.B. works for Wolfram Research, who provided the computational resources used in this work. The work herein is independent of M.B.'s employment responsibilities, and Wolfram Research had no role in it. All numerical simulations were done in Mathematica V12.2 [14].

Bibliography

[1] More background information on the logistic function: `https://en.wikipedia.org/wiki/Logistic_function`

[2] Eric W. Weisstein, "Logistic Equation." From MathWorld–A Wolfram Web Resource. `https://mathworld.wolfram.com/LogisticEquation.html`

[3] A. Ahmed and et al. "Analysis coronavirus disease (COVID-19) model using numerical approaches and logistic model." AIMS Bioengineering 7.3 (2020): 130.

[4] P. Wang, X. Zheng, J. Li, and B. Zhu, "Prediction of epidemic trends in covid-19 with logistic model and machine learning technics," Chaos, Solitons Fractals, p. 110 058, 2020.

[5] D Tátrai, Z Várallyay, COVID-19 epidemic outcome predictions based on logistic fitting and estimation of its reliability, `arXiv:2003.14160`, (2020).

[6] E. Aviv-Sharon, A. Aharoni, Generalized logistic growth modeling of the Covid-19 pandemic in Asia, Infectious Disease Modelling, 5, 502–509, 2020.

[7] S. Sakanoue, Extended logistic model for growth of single-species populations. Ecological Modelling, 205:159–68, 2007.

[8] K. Roosa and et al. Real-time forecasts of the COVID-19 epidemic in China from February 5th to February 24th, 2020. Infectious Disease Modelling, 5:256-263, 2020.

[9] N. Afshordi, B. Holder, M. Bahrami, and D. Lichtblau, "Diverse local epidemics reveal the distinct effects of population density, demographics, climate, depletion of susceptibles, and intervention in the first wave of covid-19 in the united states," arXiv:2007.00159[q-bio.PE] (2020).

[10] F. J. Richards, A Flexible Growth Function for Empirical Use. Journal of Experimental Botany 10 (2), 290–301, 1959.

[11] J. I. Steinfeld, J. S. Francisco and W. L. Hase, Chemical Kinetics and Dynamics (2nd ed., Prentice-Hall 1999) p.151-2

[12] Eric W. Weisstein, "Logistic Map." From MathWorld–A Wolfram Web Resource. https://mathworld.wolfram.com/LogisticMap.html

[13] Wolfram Research, "Epidemic Data for Novel Coronavirus COVID-19" from the Wolfram Data Repository (2021). https://doi.org/10.24097/wolfram.04123.data

[14] Wolfram Research, Inc., Mathematica, Version 12.2, Champaign, IL (2020).

[15] R. Burger, G. Chowell, and L. Y. Lara-Diiaz, Comparative analysis of phenomenological growth models applied to epidemic outbreaks. Math. Biosci. Eng. 16, 4250–4273 (2019).

Epidemic Modeling with Agent Based Models and Reinforcement Learning

Emir Arditi[1,*], Egemen Sert[2,*], Alfredo J. Morales[3,**]

[1] Ozyegin University, Istanbul, Turkey.
[2] Middle East Technical University, Ankara, Turkey.
[3] MIT Media Lab, Cambridge, MA, USA.
*Equal contribution. **Corresponding author: alfredom@mit.edu

COVID-19 evidenced that the world has transformed deeply over the last decades. We live in a more complex, volatile and uncertain world, where societies are tied to one another and interdependent. Pandemics threaten to become more severe and more frequent as the world becomes interconnected. Understanding the space of possibilities that arise from social dynamics is crucial for creating strategies and policies. We combine Agent Based Models (ABMs) with Reinforcement Learning (RL) and create simulations where agents decide on strategies to follow under varied conditions using Artificial Intelligence (AI). We apply our approach to the Susceptible-Infected (SI) epidemic model. Across multiple simulations, agents are able to control the epidemic by spontaneously clustering themselves by health, and keeping distance from the infected ones. Moreover, healthy agents implement self-isolation as the transmission probability increases, perhaps due to the risk of exposure in comparison with clustered populations. These results show that incorporating RL to ABM unveils possible strategies that are otherwise difficult to find by random exploration.

Introduction

We live in a complex, volatile and uncertain world [1]. Pandemics threaten to become more frequent as societies are tied to one another and highly interdependent. Institutions and organizations cannot manage such complexity [2]. Their methods are not conditional on the natural behavior of social systems and therefore become incapable of dealing with them effectively, imposing fragile strategies upon the population at large [3]. While there has been a large body of research around the mechanistic behavior of epidemics across social networks and environments, there is not a clear consensus on how to respond to epidemics at the scale of population. In this regard, it is imperative to explore and analyze the behavior of models from multiple perspectives, including irreducible behaviors that emerge from fundamental interaction mechanisms and utilitarian structures. The combination of Agent Based Modeling (ABM) and Reinforcement Learning (RL) provides the opportunity to explore the space of possible responses for social phenomena such as segregation, polarization, or spreading of information [4]. RL provides agents with artificial intelligence, such that their actions are decided following certain policies and objectives under multiple environmental conditions. While ABM shows the complexity of social processes and emergent patterns, RL provides possibilities and strategies to operate under such conditions.

Agent Based Modeling (ABM) is a simulation framework to study the behavior of biological and social systems [5] based on the interactions among individuals or agents [6]. The models show that macroscopic behaviors and regularities emerge from the aggregation of multiple distributed interactions and decisions [7] with capabilities to map the space of possibilities [8]. ABM tests theories against simulations [9] with emphasis on heterogeneous, autonomous actors operating with incomplete information [10]. They have been applied to study economic systems [11, 12], as well as individual [13] and organizational [14, 15] decision making processes, including the design of distributed and autonomous systems such as traffic control [16] and energy management [17]. Applications to social systems include wealth distributions [9], politics [18], global systems [19], and cultures [20], among others [5]. In biological systems, ABM has shown a remarkable power to explain epidemics [21, 22], human body systems [23], ecosystems [24], and links between biology and social behaviors [25]. Despite the wide-range of applications of ABM simulations, agents often take actions blindly and have very limited capacity, if any, to evaluate the effects of their actions. With the introduction of RL, agents anticipate the effects of their actions in the future based on Markov Decision Processes, and generate strategies towards reaching goals, in a richer modeling framework.

Reinforcement Learning (RL) provides agents with the capability to learn from their experience during simulations and become aware of their environment. Agents adapt their behavior according to an interplay between a previously defined structure of rewards and the state of the environment. Multi-Agent Reinforcement Learning (MARL) is the method when multiple agents are employed. The combination of RL with Deep Neural Networks have recently achieved unprecedented human performance in complex tasks such as gaming [26], rapid motion [27], and communication with incomplete information [28]. More recently, RL has been applied to study social dynamics [29] such as segregation [4], cooperation [30, 31], and game theory [32, 33].

In the last year, researchers have developed RL solutions for epidemic control using available data. COVID-19 prevention policies, such as closing schools, have been analyzed and optimized using Great Britain's connectivity data [34]. Moreover, extensions of epidemic models include new

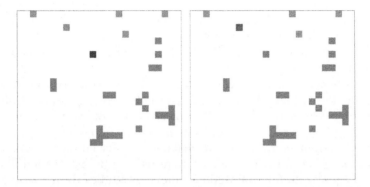

Figure 1: Example displays of two of the states of our environment (colored for clarity). In Reinforcement Learning setups, the environment is defined as the simulation of a task and state is defined as the status of the environment at time t. Our environment consists of a grid in which the agents try to stay healthy as much as possible. Here, the red color represents infected agents and the green color represents healthy agents. Darker green (left) and darker purple (right) colors indicate the agents that about to take an action and they perceive their environment as seen in the respective figures.

properties such as asymptomatic transmission [35] and the intervention of governments [36]. We introduce RL to the Susceptible-Infected (SI) model and observe how agents adapt and attempt to control the spread of the epidemic. Our results show the emergence of spontaneous separation in space or segregation between healthy and infected agents. Our methods can be extended to study other types of social phenomena and inform decision makers on possible actions.

Model

Modeling Epidemics

Modeling the spread of diseases, behaviors and information across social systems has drastically advanced our understanding of events such as pandemics, news diffusion, and malware spreading [37]. A large number of these models rely on the Susceptible - Infected - Recovered (SIR) modeling framework [38]. SIR is an ABM where agents can either be in three states: susceptible to get infected, infected or recovered from the infection. Susceptible agents can get infected from other infected individuals with a probability of transmission β. Infected agents can recover from the disease with a probability of recovery γ. The model yields various outcomes based on the parameters such as infection and recovery probabilities as well as the population density of agents. A particular example called SI occurs when the recovery probability is zero ($\gamma = 0$) and infected agents cannot get recovered. In the SI model, all agents may eventually get infected over time if agents are randomly contacting each other. Another variation includes becoming susceptible again, called the SIS models, and agents can relapse on the infection.

The behavior of the SIR-type models has been studied on multiple network topologies [39]. When applied to networks, agents can only get infected from their neighbors, which means that risks from infection vary according to one's social network rather than being homogeneous across the whole population. Therefore making an infected agent with many connections (edges) to other agents more risky to the population than an infected agent with few connections. Previous studies show that the critical mass required to trigger an epidemic decreases to zero in the limit as the network becomes complex and hubs that connect large portions of the network emerge and centralize connections [22].

In our experiments we apply the SI model on spatial networks based on simulations using RL. We adapt the rules of interactions from the traditional agent-based model to the framework of AI and RL. This allow us to introduce new rewards and observe emergent behaviors. In this model, agents (nodes) can move across free locations on a grid space and interact with those individuals within their space of observation. Edges are formed among agents that are at most two grids apart from each other. Therefore, agents create a dynamic social network from spatial interactions that is subject to change as agents move at each iteration.

We initialize the model by creating N agents on the grid space: S susceptible and I infected ($N = S + I$). At each iteration, a social network is formed based on spatial closeness. Then infected agents spread the disease to susceptible neighbors with a transmission probability β. Note that the spreading phenomena is based on connectivity of infected agents. On one hand, the disease will spread quickly if infected agents have too many connections to healthy ones. On the other hand, the infection is contained in a fixed population if infected agents have no healthy neighbors. The dynamics of the emergent social network is based on agents' actions. The collective goal is to minimize the chances of getting infected. Global patterns of behavior may emerge due to the self-organization of agents' actions. An example of these behaviors include segregation and isolation.

Multi-Agent Reinforcement Learning

The modeled learning environment consists of a grid space containing N autonomous agents. At $t = 0$, S of them are healthy and $I = N - S$ of them are infected. The goal of the environment is to contain the spread of the disease by maximizing the cumulative reward we explain in this section. The system rewards agents if they keep the population healthy, while the SI dynamics run in the background. Defining the reward function and agent state space is fundamental for designing stochastic tasks with clear goals.

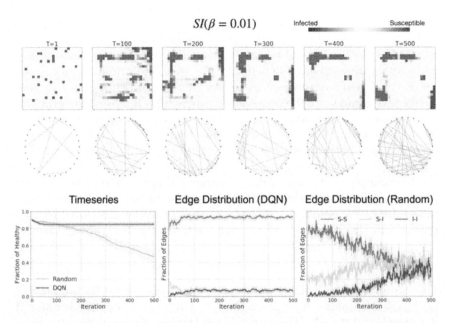

Figure 2: Average heat-map of healthy-infected agent locations (top), social network of agents per hundred iterations (middle) and dynamics of various statistics (bottom) for SI model with probability of transmission $\beta = 0.01$. Green indicates healthy agents and purple indicates the infected, legend is placed on upper right. Time series compare the performance of DQN with random action policy as a control group. On edge distribution figures, healthy-healthy edges are denoted in green (S-S), healthy-infected (risky) edges are denoted in pink (S-I), and infected-infected edges are denoted in purple (I-I). Transparent regions indicate the standard deviation of metrics estimated on ten replications of each experiment.

We created a learner model that can handle multi-agent simulations using Deep Q-Learning [40] as our primary learner due to its effectiveness on complex environments. The learner is capable of inputting a discrete state and action space and learning both how to model the spread of the infection and how to prevent it. Direct models such as classic Q-Learning or Value Iteration are not suitable for this case given the difficulties to map the state space and action space a priori.

Environment Definition

The RL model receives the whole grid space as an input. The model controls each agent's action in a one-by-one manner. The implemented state definition is as follows. For the selected agent, $s = 2$ if it is healthy and $s = -2$ otherwise. For other agents, $s = 1$ if they are healthy and $s = -1$ otherwise.

Our environment consists of 5 actions, namely: Left, Up, Right, Down and Stay. The first four imply moving to a new location. The last action is implemented since it could happen that agents must stay healthy at all cost and moving might not be beneficial. The state and action space defined for our designed environment let agents move separately, but also defines a collective mind since the observation comes from the complete grid. We define states for each location based on their type of occupation. If the location is occupied by an infected agent then $s = -1$. If the location is occupied by a healthy agent then $s = 1$. Otherwise, $s = 0$ if the location is empty.

An example of the simulation environment can be found on Figure 1 where we show the location of multiple healthy (green) and infected (red) individuals. We emphasise the selection of a healthy individual (darker green) in the left panel and an infected one (darker red) in the right panel.

Reward Function

The cumulative goal of the environment is to maximize the number of healthy individuals. Agents receive $r = 1$ reward if they are healthy at iteration t, and $r = -1$ if they are infected. Unlike the usual RL scenario, a separate terminal punishment or reward is not implemented. The reward function is evaluated at the selected agent and it enables the neural network to apply self-importance on each agent separately.

Network Architecture

The Neural Network (NN) of our DQN (Deep Q-Network) model is a simple Convolutional Neural Network (CNN), containing 3 CNN layers and a single fully connected layer. The network inputs the whole grid and returns a 5 dimensional vector of actions for each agent. Each value shows the remaining cumulative reward estimate, $Q(s, a)$, for each action in the current state. Adam optimizer [41] is used with a learning rate of 10^{-4} for applying batch gradient descent. The neural network is trained from a batch of random samples with a batch size of 32. In order to stabilize the learning process, the target Q-values are taken from a target network. After each 800 steps, the target network is updated with the current parameters of the main network [40].

Experiment Details

We use a grid world of 24x24 locations in our experiments. The reason for this selection is due to the computational complexity of our CNN model. There are $N = 30$ agents inside the grid. At $t = 0$, 10% percent of the population is initially infected. The disease can transmit within a radius of 2 locations from each agent. A sick person can only make another person sick if the distance between them is 2 locations or less. The experimental

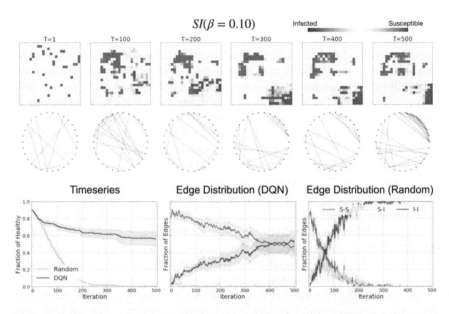

Figure 3: Average heat-map of healthy-infected agent locations (top), social network of agents per hundred iterations (middle) and dynamics of various statistics (bottom) for SI model with probability of transmission $\beta = 0.10$. Green indicates healthy agents and purple indicates the infected, legend is placed on upper right. Timeseries compare the performance of DQN with random action policy as a control group. On edge distribution figures, healthy-healthy edges are denoted in green (S-S), healthy-infected (risky) edges are denoted in pink (S-I), and infected-infected edges are denoted in purple (I-I). Transparent regions indicate the standard deviation of metrics estimated on ten replications of each experiment.

setup is repeated for different transmission probabilities β in order to study their effects on the population behavior.

Simulation steps are individually defined as N state-action pairs, where N indicates the number of agents in the system. This is different from traditional implementations where $N = 1$. We choose $N > 1$ because the neural network processes all agents inside the system separately and outputs a different action for each agent. At the start of each step, the list of agents is shuffled and the current states for each agent is passed to the system one-by-one. A regular episode takes 500 steps during training, but episodes terminate early if all agents become either healthy or infected. For each infection probability, β, training consists of 200 episodes. Later, the current model and a random agent is tested for 10 more episodes and their results are compared to one another. The environment and simulations have been implemented using Python Libraries and are available for download.[1]

---

[1]Link to code: https://github.com/emirarditi/EpidemicModelingWithABMandRL

Results

We inspect the dynamics of our model under three different transmission probabilities: $\beta = 0.01$, $\beta = 0.10$ and $\beta = 0.90$. For each transmission probability, we train the DQN environment until we see convergence and tested its capacity to reduce or contain the transmission of the disease in 10 different realizations. We measure the significance of the results by comparing the performance of DQN environments with simulations from control groups where agents are limited to random actions. The random realizations consist of agents moving randomly in the grid space, without using any of the RL capabilities and being unaware of their environment. The initialization of the environment in these simulations are consistent with the previous 10 realizations.

The results obtained with transmission probability $\beta = 0.01$ are shown in Figure 2. The top row shows heat-maps with the average of spatial locations of healthy and infected agents over the last 100 iterations. Dominantly healthy areas are represented in green and infected ones in purple. The maps indicate that agents learn to isolate and cluster sick individuals and separate the space into healthy and infected regions. The middle row shows the spatial networks of agents at time t. Nodes represent agents and connections indicate close proximity in space. The color of the edges is consistent with the health of the individual represented as a node. The connections are segregated across the network according to the health of the individual.

The bottom row shows the performance of various metrics as the simulations evolve. The results show that DQN environments learn to create safe spaces by segregating the interactions between healthy and sick individuals. The bottom left panel shows the performance of DQN with respect to random control group in terms of fraction of population that remains healthy over time. The healthy population in DQN environments (blue) remains consistently higher than the random case (yellow) which decreases over time. The bottom middle panel shows the dynamics of fraction of edges between various edge types when DQN policy is employed. Edge types are as following: healthy-healthy agents (S-S, green), healthy-infected agents (S-I, pink), and infected-infected agents (I-I, purple). Finally, the bottom right panel shows the dynamics of fraction of edges when random policy is employed with the same color code. The segregation of interactions by health is consistently higher in DQN environments. Solid lines indicate average behavior after 10 realizations. Transparent region indicates one standard deviation span in performance.

Consistent results are presented in Figures 3 and 4 for other values of transmission probabilities $\beta = 0.10$ and $\beta = 0.90$ respectively. These figures show the behavior of the system when the infection severity and transmission probabilities increase. In Figure 3, if a healthy agent is a neighbor of an infected agent, it gets infected with 10% probability. Moreover, in Figure

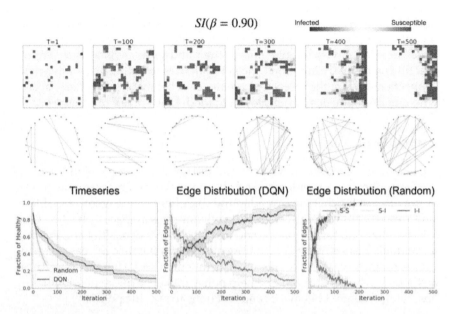

Figure 4: Average heat-map of healthy-infected agent locations (top), social network of agents per hundred iterations (middle) and dynamics of various statistics (bottom) for SI model with probability of transmission $\beta = 0.90$. Green indicates healthy agents and purple indicates the infected, legend is placed on upper right. Timeseries compare the performance of DQN with random action policy as a control group. On edge distribution figures, healthy-healthy edges are denoted in green (S-S), healthy-infected (risky) edges are denoted in pink (S-I), and infected-infected edges are denoted in purple (I-I). Transparent regions indicate the standard deviation of metrics estimated on ten replications of each experiment.

4, the agent gets infected with 90% probability, which is an almost certain event. In both settings, the DQN environment outperforms the random policy and manages to keep a portion of the population healthy. Moreover, we see the emergence of social distancing as a new type of behavior. As the severity of the disease increases, the number of edges between healthy-healthy agents decreases which means that they are not staying in close proximity to each other despite being both apparently healthy. Moreover, in the extreme case of $\beta = 0.90$, agents choose self-isolation, despite not being explicitly encouraged. We believe that this behavior is due to connectedness of healthy agents increasing the risk of collective infection as the severity of the disease increases.

The results shown in the bottom panels of Figures 2 and 4 indicate that DQN environments effectively learn to contain the disease, despite not having a probability of recovery. This observation contributes to the traditional SI modeling framework which shows that single-component networks will get infected over time. The addition of AI and RL shows that agents can alter the structure of their social network so that isolation from the infection is granted and diseases controlled. Finally, the distribution of edges in DQN environments hints at the mindset of DQN agents. We see that during the early iterations, the network tries to minimize healthy-risky connections so that the disease is contained. A behavior that is fairly absent when random policies are employed to model the spread of diseases.

Conclusion

We combine Agent-Based Modeling (ABM) with Reinforcement Learning (RL) in application to epidemics and observe if agents can control the spread of the disease by changing or adapting their behavior. Agents are able to slow down the epidemic by segregating themselves in space by health. They self-organize and form separate and distant clusters of healthy and infected individuals respectively. As the transmission probability increases, agents choose to self-isolate and reduce even further their social interactions. These methods can be generalized and help policy makers explore possibilities by observing emergent behaviors and reactions to specific policies.

Bibliography

[1] J. Balsa-Barreiro, A. Vie, A. Morales, and M. Cebrian, "Deglobalization in a hyper-connected world.," *Palgrave Communications*, vol. 6, no. 28, 2020.

[2] J. C. Scott, *Seeing like a state: How certain schemes to improve the human condition have failed*. Yale University Press, 2020.

[3] N. N. Taleb, *Antifragile: Things that gain from disorder*, vol. 3. Random House Incorporated, 2012.

[4] E. Sert, Y. Bar-Yam, and A. J. Morales, "Segregation dynamics with reinforcement learning and agent based modeling," *Scientific reports*, vol. 10, no. 1, pp. 1–12, 2020.

[5] C. M. Macal and M. J. North, "Agent-based modeling and simulation," in *Proceedings of the 2009 Winter Simulation Conference (WSC)*, pp. 86–98, IEEE, 2009.

[6] H. Sayama, *Introduction to the modeling and analysis of complex systems*. Open SUNY Textbooks, 2015.

[7] T. C. Schelling, "Dynamic models of segregation," *Journal of mathematical sociology*, vol. 1, no. 2, pp. 143–186, 1971.

[8] S. Hassan, J. Arroyo, J. M. Galán, L. Antunes, and J. Pavón, "Asking the oracle: Introducing forecasting principles into agent-based modelling," *Journal of Artificial Societies and Social Simulation*, vol. 16, no. 3, p. 13, 2013.

[9] J. M. Epstein and R. Axtell, *Growing artificial societies: social science from the bottom up*. Brookings Institution Press, 1996.

[10] J. M. Epstein, "Agent-based computational models and generative social science," *Complexity*, vol. 4, no. 5, pp. 41–60, 1999.

[11] H. Kita, K. Taniguchi, and Y. Nakajima, *Realistic Simulation of Financial Markets: Analyzing Market Behaviors by the Third Mode of Science*, vol. 4. Springer, 2016.

[12] M. Oldham, "Introducing a multi-asset stock market to test the power of investor networks," *Journal of Artificial Societies and Social Simulation*, vol. 20, no. 4, p. 13, 2017.

[13] T. Balke and N. Gilbert, "How do agents make decisions? a survey," *Journal of Artificial Societies and Social Simulation*, vol. 17, no. 4, p. 13, 2014.

[14] W.-S. Yun, I.-C. Moon, and T.-E. Lee, "Agent-based simulation of time to decide: Military commands and time delays," *Journal of Artificial Societies and Social Simulation*, vol. 18, no. 4, p. 10, 2015.

[15] K. H. van Dam, Z. Lukszo, L. Ferreira, and A. Sirikijpanichkul, "Planning the location of intermodal freight hubs: an agent based approach," in *2007 IEEE International Conference on Networking, Sensing and Control*, pp. 187–192, April 2007.

[16] S. Kumar and S. Mitra, "Self-organizing traffic at a malfunctioning intersection," *Journal of Artificial Societies and Social Simulation*, vol. 9, no. 4, p. 3, 2006.

[17] T. Ma and Y. Nakamori, "Modeling technological change in energy systems–from optimization to agent-based modeling," *Energy*, vol. 34, no. 7, pp. 873–879, 2009.

[18] R. Axelrod, "A model of the emergence of new political actors," in *Artificial Societies*, pp. 27–44, Routledge, 2006.

[19] L.-E. Cederman, *Emergent actors in world politics: how states and nations develop and dissolve*, vol. 2. Princeton University Press, 1997.

[20] R. Axelrod, "The dissemination of culture: A model with local convergence and global polarization," *Journal of conflict resolution*, vol. 41, no. 2, pp. 203–226, 1997.

[21] V. Wong, D. Cooney, and Y. Bar-Yam, "Beyond contact tracing: community-based early detection for ebola response," *PLoS currents*, vol. 8, 2016.

[22] R. Pastor-Satorras, C. Castellano, P. Van Mieghem, and A. Vespignani, "Epidemic processes in complex networks," *Reviews of modern physics*, vol. 87, no. 3, p. 925, 2015.

[23] V. A. Folcik, G. C. An, and C. G. Orosz, "The basic immune simulator: an agent-based model to study the interactions between innate and adaptive immunity," *Theoretical Biology and Medical Modelling*, vol. 4, no. 1, p. 39, 2007.

[24] E. M. Rauch and Y. Bar-Yam, "Long-range interactions and evolutionary stability in a predator-prey system," *Phys. Rev. E*, vol. 73, p. 020903, Feb 2006.

[25] M. Hartshorn, A. Kaznatcheev, and T. Shultz, "The evolutionary dominance of ethnocentric cooperation," *Journal of Artificial Societies and Social Simulation*, vol. 16, no. 3, p. 7, 2013.

[26] V. Mnih, K. Kavukcuoglu, D. Silver, A. A. Rusu, J. Veness, M. G. Bellemare, A. Graves, M. Riedmiller, A. K. Fidjeland, G. Ostrovski, *et al.*, "Human-level control through deep reinforcement learning," *Nature*, vol. 518, no. 7540, p. 529, 2015.

[27] N. Heess, S. Sriram, J. Lemmon, J. Merel, G. Wayne, Y. Tassa, T. Erez, Z. Wang, S. Eslami, M. Riedmiller, *et al.*, "Emergence of locomotion behaviours in rich environments," *arXiv preprint arXiv:1707.02286*, 2017.

[28] E. Sert, C. Sönmez, S. Baghaee, and E. Uysal-Biyikoglu, "Optimizing age of information on real-life tcp/ip connections through reinforcement learning," in *2018 26th Signal Processing and Communications Applications Conference (SIU)*, pp. 1–4, IEEE, 2018.

[29] M. Lanctot, V. Zambaldi, A. Gruslys, A. Lazaridou, K. Tuyls, J. Pérolat, D. Silver, and T. Graepel, "A unified game-theoretic approach to multiagent reinforcement learning," in *Advances in Neural Information Processing Systems*, pp. 4190–4203, 2017.

[30] E. M. de Cote, A. Lazaric, and M. Restelli, "Learning to cooperate in multiagent social dilemmas," in *Proceedings of the fifth international joint conference on Autonomous agents and multiagent systems*, pp. 783–785, ACM, 2006.

[31] J. Z. Leibo, V. Zambaldi, M. Lanctot, J. Marecki, and T. Graepel, "Multi-agent reinforcement learning in sequential social dilemmas," in *Proceedings of the 16th Conference on Autonomous Agents and MultiAgent Systems*, pp. 464–473, International Foundation for Autonomous Agents and Multiagent Systems, 2017.

[32] T. W. Sandholm and R. H. Crites, "Multiagent reinforcement learning in the iterated prisoner's dilemma," *Biosystems*, vol. 37, no. 1-2, pp. 147–166, 1996.

[33] M. Wunder, M. L. Littman, and M. Babes, "Classes of multiagent q-learning dynamics with epsilon-greedy exploration," in *Proceedings of the 27th International Conference on Machine Learning (ICML-10)*, pp. 1167–1174, Citeseer, 2010.

[34] P. Libin, A. Moonens, T. Verstraeten, F. Perez-Sanjines, N. Hens, P. Lemey, and A. Nowé, "Deep reinforcement learning for large-scale epidemic control," *arXiv preprint arXiv:2003.13676*, 2020.

[35] H. Khadilkar, T. Ganu, and D. P. Seetharam, "Optimising lockdown policies for epidemic control using reinforcement learning," *Transactions of the Indian National Academy of Engineering*, vol. 5, no. 2, pp. 129–132, 2020.

[36] V. Kompella, R. Capobianco, S. Jong, J. Browne, S. Fox, L. Meyers, P. Wurman, and P. Stone, "Reinforcement learning for optimization of covid-19 mitigation policies," *arXiv preprint arXiv:2010.10560*, 2020.

[37] A.-L. Barabási, "Network science," *Philosophical Transactions of the Royal Society A: Mathematical, Physical and Engineering Sciences*, vol. 371, no. 1987, p. 20120375, 2013.

[38] W. O. Kermack and A. G. McKendrick, "Contributions to the mathematical theory of epidemics—i," *Bulletin of mathematical biology*, vol. 53, no. 1-2, pp. 33–55, 1991.

[39] M. E. J. Newman, "Spread of epidemic disease on networks," *Phys. Rev. E*, vol. 66, p. 016128, Jul 2002.

[40] V. Mnih, K. Kavukcuoglu, D. Silver, A. A. Rusu, J. Veness, M. G. Bellemare, A. Graves, M. Riedmiller, A. K. Fidjeland, G. Ostrovski, S. Petersen, C. Beattie, A. Sadik, I. Antonoglou, H. King, D. Kumaran, D. Wierstra, S. Legg, and D. Hassabis, "Human-level control through deep reinforcement learning," *Nature*, vol. 518, pp. 529–533, Feb. 2015.

[41] D. Kingma and J. Ba, "Adam: A method for stochastic optimization," *International Conference on Learning Representations*, 12 2014.

Life, Liberty, and Lockdowns

Philip Z. Maymin[1,*] and Zakhar G. Maymin

[1]Dolan School of Business, Fairfield University, Fairfield, CT, USA.
*Corresponding author: pmaymin@fairfield.edu

We offer a computational model for evaluating tradeoffs between life and progress. Our framework evaluates people as cells in a one-dimensional outer-totalistic cellular automaton with two living states and an absorbing output state of death. Pandemics are modelled as instant death for a portion of the population and government regulations are modelled as lockdown restrictions on the number of consecutive neighboring cells in a certain state. With this model, we are able to compare the implicit trade-off between an unchecked yet instant pandemic and a continual governmental lockdown. We find that lockdowns lead to reduced complexity and increased death compared to a pandemic. If people are allowed to vote, they tend to vote for lockdowns early but regret their choice later. The findings suggest generally that societies can be robust to external attacks but can wither from internal attempts to control the mechanisms of progress.

1 Introduction

There are two fundamentally different ways of modeling phenomenon: from the top down and from the bottom up.

Top-down approaches start with stylized facts and calibrate models with continuous parameters to match those stylized facts as closely as possible. Bottom-up approaches first aim to simplify the problem as much as possible and then explore the resulting computational universe. Examples of such bottom-up approaches in finance and economics include [1], [2], [3], and [4].

Top-down approaches are the contemporary dominant standard in the scientific literature, even to the extent of how papers are structured, with claims outlined first, then a method, results, and discussion. However, per Wolfram's Principle of Computational Irreducibility [5], a bottom-up approach can never in principle be expressed in such a way, because it would be impossible to know ahead of time the results of an arbitrary computation. Therefore, this paper is organized in a way that may appear less conventional but is more appropriate to the computational exploration approach.

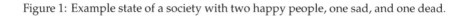

Figure 1: Example state of a society with two happy people, one sad, and one dead.

Section 2 argues for, models, and finds the unique minimal computational model of society. Section 3 applies that model to a world where a pandemic can instantly kill some of the participants, or a lockdown can prohibit certain kinds of social interactions. Section 4 concludes with suggestions for future research.

2   The Minimal Model of Society

We model the state of society as a cyclical list of people. Each person can be happy, sad, or dead. Death is an absorbing state. These can also be interpreted as healthy/susceptible, infected, and dead in the standard SIR model (e.g. [6]).

We color the three states as happy $\rightarrow$ black, sad $\rightarrow$ white, and dead $\rightarrow$ red. Fig. 1 shows an example state of two happy people, one sad, and one dead.

Each person transitions to a new state depending on their own current state and the states of their nearest living neighbors. Dead neighbors are ignored.

Ignoring dead neighbors effectively shrinks a society, but we keep the strands of red to visualize death over time. People care about how their living neighbors are doing, whether they are happy or sad, but they only care about the total. It doesn't matter if it's your left-neighbor who is sad and your right-neighbor who is happy or vice versa. This is commonly referred to as an "outer-totalistic" rule.

However, it is not a standard outer totalistic rule because there are two input colors but three output colors: a dead person does not evolve, but a living person can be either sad or happy and can become either sad, happy, or dead.

How many distinct initial states are there for each person, assuming a neighborhood region or radius of r? Temporarily renumber the states as zero for sad and one for happy in order to count the number of possible totals. Then the person evolving can be in one of two states, and the total of his $2r$ neighbors can be anywhere from zero to $2r$, which is $2r+1$ possibilities. That's $2 \cdot (2r + 1)$ possible initial states for each person.

How many different rules are there? Each of those possible initial states can be mapped onto one of three outputs, so there are $3^{2 \cdot (2r+1)}$ possible rules. Table 1 computes these maximums for radii from one to five.

Thus, there are 729 one-neighbor outer-totalistic rules from two-color inputs to three-color outputs. With a standard mapping from integer rule numbers to transition rules, Fig. 2 shows, for example, that according to rule

Figure 2: The rule plot for rule 92, showing the possible neighbors in the first row with the current cell in the middle, and the result in the middle of the second row.

Figure 3: The evolution of rule 92 for two time steps starting from the initial happy, happy, sad, dead condition.

92, a sad person surrounded by one happy neighbor and one sad neighbor will evolve to be happy in the next step, regardless of which neighbor is which.

In fact, rule 92 always results in a happy black cell if all three cells are happy or if exactly one of the three is happy and the others sad. If nobody is happy, or two people are happy and one is left out, then the cell becomes sad.

We can repeatedly apply this rule on every group of neighboring cells to generate the state of the society at the next time step , and then repeat to generate an evolution of society over time.

Fig. 3 shows the evolution of the happy, happy, sad, dead state according to rule 92 for two time steps. Everyone who was happy becomes sad, the sad person stays sad, and dead people remain dead.

Notice that for the given initial condition, rule 92 cycled after two steps. From step 2 to step 3, the evolution does nothing, so all future steps would look the same: three sad people and one dead one.

Our initial conditions for the present analysis will always be a fixed number of people, rather than a constantly growing population. Therefore, every rule will eventually cycle, some faster than others.

r	MaxRules[r]
1	729
2	59049
3	4782969
4	387420489
5	31381059609

Table 1: Number of possible rules for a given neighborhood region or radius r, where MaxRules[r_] := 3^(2 (2 r + 1)).

We typically wish to filter out cyclical evolutions and explore the space of possible evolutions, ignoring simple equivalencies such as rules that are identical except the identification of sad and happy.

Starting with a standard initial condition of one happy person in the middle of one hundred sad people, we can explore all of the possible 729 rules for emergent complexity.

While we define complexity more specifically below, for now without loss of generality we select only those evolutions that have the maximal length before they cycle.

Fig. 4 shows all these maximally complex radius-1 rules. Only fifteen inequivalent rules evolved for 500 time steps without cycling.

There is some interesting structure but the evolution is mainly symmetric. Rules 103, 148, and 263 are somewhat ironic as the first happy person dies immediately, and that is the only death that ever occurs.

What if we start with the asymmetric initial condition shown in Fig. 5?

Fig. 6 shows all the maximally complex rules starting from the asymmetric condition. Again only fifteen nonequivalent rules evolved without cycling.

These evolutions appear far more interesting and complex.

Compare these to the evolution starting from a random initial state with the same number of people shown in Fig. 7.

Most of the complex rules are common across both initial conditions, suggesting both that our asymmetric initial condition might be sufficient for evaluating the rules, and that the rules themselves exhibit a consistency in terms of complexity.

Twelve rules are common to the fifteen asymmetric complex rules and the seventeen random complex rules.

Measures of Complexity

Complexity can be measured in one of two ways. The complexity pictured in the figures above was a time-series complexity: given a rule, evolve the society, and evaluate the complexity of the resulting evolution.

An alternative measure of complexity is cross-sectional complexity: for each possible rule, evolve all possible initial conditions by one step, and count how many distinct output states they have, and rank the rules by that number. This measures how complex a rule is relative to other rules.

In our case, there are too many possible initial conditions to do an exhaustive search, so to estimate the consistency of each rule's cross-sectional complexity across different lengths of initial conditions, we can sample randomly from all of the 2^n possible initial conditions for a population size n, evolve each one for one time step, and see how many distinct output states are generated. The more distinct output states there are, the more complex that rule is for that population level. For example, a rule that always kills everyone would be minimally complex.

Figure 4: All maximally complex radius-1 rules for the standard initial condition of one happy person in the middle of 100 sad people, evolved up to 500 time steps.

Figure 5: An asymmetric initial condition: one sad, two happy, three sad, and so on.

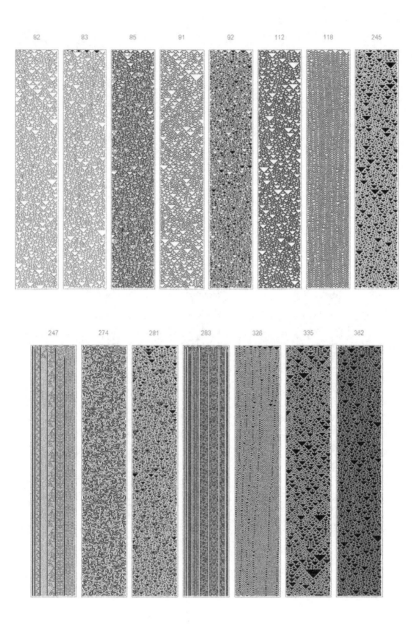

Figure 6: All maximally complex rules for the asymmetric initial condition.

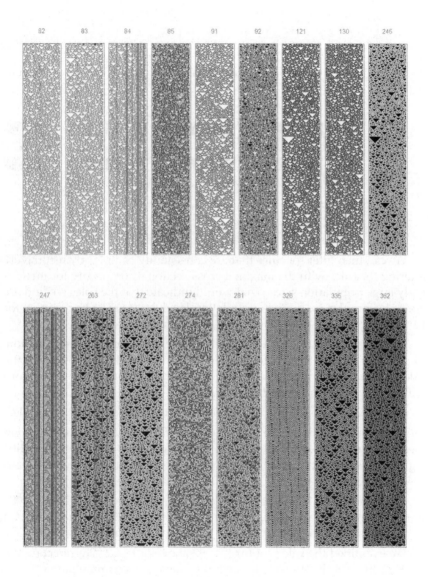

Figure 7: All maximally complex rules for a random initial condition.

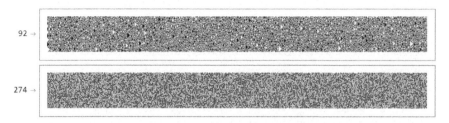

Figure 8: The only two maximally and consistently complex rules for 1000 time steps.

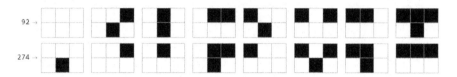

Figure 9: The rule plots for rules 92 and 274.

We can then filter for rules that are consistently complex by recursively filtering for rules with the maximal cross-sectional complexity for increasingly large population sizes. That would indicate that the rules we find are not only complex for a given population level, but tend to be complex for various population levels; thus, they are intrinsically complex.

Specifically, we start with a population of 5, evolve every rule one time step from each of 1000 different random starting conditions, and keep the rules with maximal distinct outputs. Recursively filter further while doubling the population size to 10, 20, 40, and 80.

Of the 729 possible rules, 144 are consistently cross-sectionally complex.

Let's evolve them on our asymmetric starting condition and see what they look like, transposed because of their thinness for better visibility. Fig. 8 shows the result. Only two maximally complex rules remain: 92 and 274.

Note that this is a subset of all the maximally complex rules for 1000 time steps only, because we pre-filtered only for those rules that maintain their characteristic complexity across a variety of population amounts.

We examined the rule plot for rule 92 above in Fig. 2. We can compare rules 92 and 274 side-by-side to see if they have any common patterns.

Fig. 9 shows that rule 92 is the exact opposite of rule 274: whenever rule 92 would evolve to a black cell, rule 274 evolves to a white cell, and vice versa.

Therefore we can without loss of generality call rule 92 the unique minimal model of society.

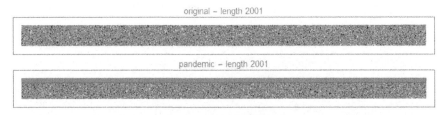

Figure 10: Evolution of the original and pandemic societies for 2000 time steps.

Figure 11: An illustration of the government prohibition on ten or more consecutive happy people.

3 Pandemic vs. Lockdown

One way to implement the effect of an unchecked pandemic is to presume that some portion of the population will die immediately. This effectively assumes that some portion of the population would remain alive after a pandemic, even without lockdowns, quarantines, social distancing, or any other changes. Equivalently, some portion of the population is deemed immune.

Suppose for concreteness the first few columns all die instantly from an unchecked pandemic; an unchecked pandemic is one that is not in any way mitigated by social distancing, masks, vaccines, lockdowns, quarantines, or any other changes in human behavior. What would the remaining evolution of rule 92, the minimal model of society, look like?

Fix a random initial condition and suppose one quarter of the population would be instantly killed. We evolve 2,000 time steps. Fig. 10 shows the result. In this case, an instant 25 percent death rate does not thwart the remaining complexity and it does not cause any further deaths.

As an alternative, consider a government intervention criminalizing happy associations above a certain threshold. For example, suppose government edicts make any sequence of ten or more happy black cells illegal. Since any government law can ultimately be enforced only by violence, for a minimal model interpretation we can implement such a policy as instantly killing any sequence of three or more black cells, i.e., by converting their state to red. Fig. 11 illustrates this policy on the asymmetric initial condition.

With a lockdown in place, we can essentially evolve the society by interspersing ordinary societal evolution with the governmental restrictions. Fig. 12 extends the earlier figure to include a comparison with the lockdown society.

Life, Liberty, and Lockdowns

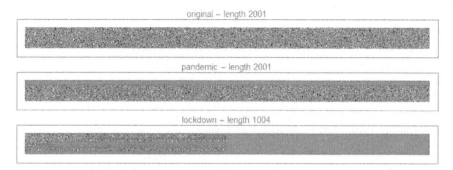

Figure 12: Evolutions of the original, pandemic, and lockdown societies.

Figure 13: One time step forecast under pandemic and lockdown scenarios, and the resulting votes. Individuals vote for the better outcome if they differ, or otherwise abstain.

In a lockdown, for both rules, fewer people die initially, but ultimately result in a total annihilation of the entire population, making the pandemic's instant but one-time population reduction of 25 percent seem utopian by comparison.

What if we allow voting? Suppose each column in the evolution is a single person who can vote either for or against government-enforced lockdowns. However, each person is cognitively or computationally limited and can only forecast their own state one row into the future. In other words, at the time of the vote, each person looks at his or her neighbors and forecasts their own cell color in the next time step. They then vote for the program that makes them happy, or at least sad but alive. In the event of a tie, they abstain.

Consider such a vote at the initial time above. The first 25 people who would be instantly killed by an unchecked pandemic would surely all vote for the lockdown, because sad or happy is better than dead. Of the remaining 75 people, one person (ironically, the 26th, the first one not to die from the instant pandemic, who would be sadder under a lockdown) would vote for the unchecked pandemic, and one person would vote for the lockdown (the last one not to die from the instant pandemic, who would be happier under a lockdown). Thus the vote would be 1-26 in favor of the lockdown, with 73 abstentions. This vote is visually summarized in Fig. 13.

Fig. 14 displays how such a comparison-based vote would look across time. Initially, and for the first few time steps, votes are overwhelmingly

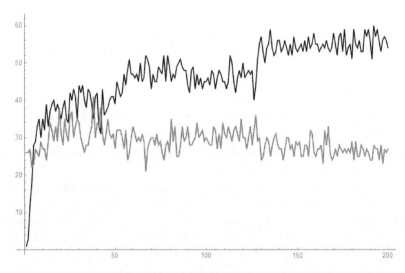

Figure 14: Vote tallies for lockdown (black) and pandemic (red) across time.

in favor of lockdown. Then the vote is contentious until about the 50th time step, after which the majority of living voters would have consistently preferred the pandemic.

4 Conclusion

Our model has effectively no parameters. It has not been calibrated to actual parameters of the Covid-19 virus. Lockdowns and other government measures aimed at reducing the spread of the pandemic are far more complicated than merely murdering adjacent happy neighbors. Most devastatingly of all, our society surely is not a one-dimensional outer-totalistic cellular automaton operating on a fixed population.

Instead, the aim of this model was to generate with the simplest possible mechanism the possible effects of government intervention vs. nonintervention. With a simplest-model approach, the goal is not to provide immediately actionable policy implications but rather to explore, illustrate, and compare counterfactual scenarios in a deterministic but computationally irreducible model.

A computationally interactive version of this paper is available.

Future extensions can explore higher radii, incorporating randomness or time delays in pandemic deaths or government regulations, allowing for changes in the evolutionary rules, and extending the voting forecast window.

As a general explanation, these illustrations and explorations suggest that a society can be automatically robust to an external attack such as a pandemic but that attempts to tweak the evolution in the name of safety may in fact make the society more fragile overall. Such government interventions will at first have widespread support but eventually people will have regretted allowing the government interventions in the first place.

Bibliography

[1] P. Z. Maymin, "Regulation simulation," *European Journal of Finance and Banking Research*, vol. 2, no. 2, p. 1–12, 2009.

[2] P. Z. Maymin, "The minimal model of financial complexity," *Quantitative Finance*, vol. 11, no. 9, p. 1371–1378, 2011.

[3] P. Z. Maymin, "A new kind of finance," *Emergence, Complexity and Computation Irreducibility and Computational Equivalence*, p. 89–99, 2013.

[4] P. Z. Maymin, "A new algorithmic approach to entangled political economy: Insights from the simplest models of complexity," *Entangled Political Economy Advances in Austrian Economics*, p. 213–236, 2014.

[5] S. Wolfram, *A New Kind of Science*. Wolfram Media, 2002.

[6] E. W. Weisstein, "SIR model," *MathWorld–A Wolfram Web Resource*.

Repetitive Rapid Testing Model for COVID-19

Diego Zviovich*

*Corresponding author: papers@zviovich.net

Non-pharmaceutical interventions (NPIs) such as testing, contact tracing, public use of face masks and other personal protective equipment
(PPE), curtailing of super-spreader events can be effective tools to keep
the spread of COVID-19 at bay. A repetitive rapid testing protocol is
explored as a public health strategy. An agent based model is used
to analyze its viability and its response to changes in parameters and
conditions.

Keywords: Agent based modeling, antigen testing, cellular automata,
COVID-19, PCR testing, rapid testing.

1   Introduction

During an outbreak of contagious disease that shows the potential for systemic harm,[1] there may not be known pharmaceutical agents (i.e. ther-
apeutics and vaccines) that are sufficient for mitigating the spread of the
disease or its physiological impact, and therefore the damage it causes to
individuals and society.

Moreover, there is no way to know a priori the time horizon on which
such pharmaceuticals may be discovered or developed. Consequently, non-
pharmaceutical interventions (NPIs) are key tools in limiting the spread
and inducing the decay of contagious pathogens, and ideally should be
sufficient for managing and extinguishing an outbreak in the absence of
effective pharmaceuticals.

NPIs ultimately aim to constrain interactions between individuals such
that those who are contagious become unlikely to amplify or spread a
pathogen in aggregate. This has manifested in various forms historically
both within and across outbreaks. For instance during the ongoing COVID-
19 pandemic alone we have observed, non-exhaustively: general travel
restrictions and travel-associated quarantines, reduction in normal societal
activities (e.g. business closures and event cancellations), PPE usage such
as masks, sanitary cordons, forced isolation of individuals, and so-called

---

[1]We operationally define systemic harm as harm that would be damaging at the societal
scale by inducing insufficiency of essential resources e.g. hospital overruns or food supply
shortages or harm that is otherwise deemed to represent an intolerable increase in likely
personal harm to a large set of individuals.

"lockdowns" which aim to limit the mobility and behavior of all individuals across some region to a greater or lesser degree [1, 2].

Without the ability to test individuals for infection, NPIs must either be applied uniformly, impacting both the infected and uninfected alike, or otherwise depend on detection of symptoms which are unreliable as a sign of infection and contagiousness. Hence, testing directly for infection ought to be viewed as a means of modifying and enabling NPI tactics, and not only as a way of tracking the progression of an outbreak or verifying the existence of individual symptomatic cases.

The development of cheap, rapid tests opens the door to a new non-pharmaceutical intervention consisting of massive, frequent testing through-out the population. These kits costing no more than US$5 do not require the assistance of a health worker. The user can obtain the result in fifteen minutes. Used in combination with a mobile application, the individual can obtain a temporary digital health pass that can be validated by third parties (workplaces, churches, schools, etc).

This paper will explore the validity of a massive high frequency testing strategy (which we'll refer as the 'strategy') in a population via agent-based modeling. We'll investigate the viability of this strategy to changes in the parameters of the disease, the testing protocol and the sensitivity and specificity of the tests.

Section 2 describes the types of tests that are available to detect if an individual is infected by the COVID-19 virus. Section 3 provides the details of the agent-based model used to simulate the results of implementing the strategy on a population. Section 4 will cover in detail the algorithm used to implement the model. Section 5 will analyze the results obtained through the execution of the model with a focus on understanding how the different parameters affect the success of the strategy and will share our conclusions regarding the viability of this strategy as a public health option.

2   Types of diagnostic tests available

There are two different categories of tests for the COVID-19: antibody tests and diagnostic tests.

Antibody tests look for antibodies generated by the immune system upon exposure to the COVID-19 virus. As these antibodies are a response to the infection of the pathogen, this type of tests measures the presence of the virus indirectly. It can take several days, up to weeks, for the antibodies to reach levels above the threshold of detection. Therefore these tests are not used to diagnose COVID-19, but to assess the individual's immunity to the virus.

Diagnostic tests are used to determined if the individual is undergo-ing an active viral infection. As of January 2021, there are two types of diagnostic tests approved in the market: molecular tests and antigen tests.

Molecular tests

The molecular tests identify a part of the viral genome in the respiratory tract specimens. The current gold standard for COVID-19 diagnosis is the reverse transcriptase-polymerase change reaction (RT-PCR). The test detects amplified COVID-19 genome residing in the specimen. This test can take between 4 and 10 hours from sample to result, but due to the logistics of testing execution, which requires specialized equipment found in major laboratories or well-equipped hospitals, it takes approximately 3-5 days to be reported back to the patients. The current cost of this test is between US$60 and $300. RT-PCR works by amplifying specific COVID-19 genomic sequence(s). Viral RNA is extracted from the patient's specimen and is purified. This RNA is then converted into a cDNA (complementary DNA) by reverse transcriptase. It is the cDNA that is subsequently amplified by the PCR. TaqMan probes are used to quantify the RNA copies by producing a fluorescence signal during the amplification cycles [3]. There are two additional molecular test technologies under development: isothermal amplification (LAMP) and CRISPR-based tests. Both are in late stages of development will be in the market soon.

Searching for information on the type I error (false positive), the median error reported between 2.3% and 5%. Regarding type II error (false negative), the literature reports a 67% median 4 days after the infection, drops to 38% at the beginning of the symptoms with the low 20% reached 3 days after the onset (approximately 8 days after the onset of symptoms) [4].

Antigen tests

Antigen tests also work by taking nasal or nasopharyngeal specimens in order to detect specific proteins from the virus. These rapid antigen tests (RATs) do not require specific and expensive machinery and can provide results in less than 30 minutes. These tests tend to be less sensitive than the molecular tests discussed above, but can be massively produced at a lower cost than the RT-PCR.

On August 26th 2020, the US FDA issued an Emergency Use Authorization for a $5, 15 minute, Covid-19 antigen test, that requires no instrumentation and can be self-administered by the patient.This type of test can be paired with a complementary phone app, allowing the user to display their test results with other people or organizations such as conventions, churches, workplaces and schools. The rapid test exhibits a demonstrated sensitivity[2] of 97.1% and specificity[3] of 98.5% in clinical study. In the literature, sensitivity is also referred as positive agreement while specificity is known as negative agreement. This will allow us to calculate the false negative error in the rapid test. False positive for the RT-PCR test is estimated

---

[2]Sensitivity measures the proportion of positives correctly identified.
[3]Specificity measures the proportion of negatives correctly identified.

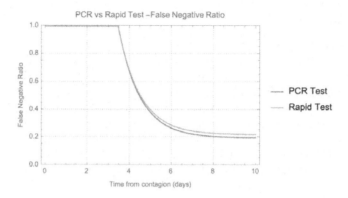

Figure 1: False negative ratio drops to a floor of 20% after 8 days from contagion. The false negative in the rapid test is estimated to be 22.1% [4].

to be 7.3%. Given that the specificity is 98.5%, we can estimate the rapid test false positive rate to be approximately 8.7% [5].

3   Exploring the different scenarios with agent based modeling

Model description

The availability of a rapid test that can be performed by any individual without the intervention of a health worker opens the possibility to allow for events or locations where people gather. By performing the test, and certifying through a mobile application that the user has no detectable antigens, the person might then be allow to engage with others in a close space event.

The scenario described above would require the following protocol:

- An individual would submit to a rapid test to determine the presence of antigens on a periodic basis.

- In the case that the test returns a positive result, the individual would move to isolation and will perform a detailed RT-PCR slow test to determine if the individual remains infectious or not (the rapid tests shows only if the person was exposed to the virus).

- People fully recovered will not need to be submitted for further testing. The CDC does no longer recommend a test-based strategy for discontinuing the isolation of most patients [6].

- It is expected that a certain ratio of individuals will break the protocol outlined above: a fraction of individuals that will avoid being tested, a different rate of will not comply with the quarantine mandate.

Compartments

Epidemiological modeling splits the host population into different classes based on their exposure to the pathogen and actions taken by the agents [9].The hosts are grouped into different compartments:

- *Susceptibles (S)*: the host is not infected but could become infected when in contact with an infected host.

- *Susceptibles Quarantined (SQ)*: host is not infected and is isolated from the rest of the population.

- *Exposed (E)*: also known as latent infection. The host has contracted the pathogen but cannot transmit it yet. The host does not exhibit any disease symptoms.

- *Exposed Quarantined (EQ)*: exposed host that is isolated from the population.

- *Infectious Asymptomatic (IA)*: the host has a high pathogen load and can transmit the pathogen to other agents. Host does not exhibit any symptoms of the disease. Members of this compartment interact with susceptible agents as under a normal pattern of behavior. Members of this compartment can potentially transmit the disease to susceptible hosts in the newtwork.

- *Infectious Symptomatic (IS)*: the host has a high pathogen load and can transmit the pathogen to other agents. Host exhibits symptoms of the disease. Agents are not under quarantine being a potential vector of contagion to susceptible agents.

- *Infectious Asymptomatic Quarantined (IAQ)*: infectious asymptomatic agents isolated from the rest of the population. These agents are under quarantine restrictions. Their isolation reduces the probability of infecting susceptible agents.

- *Infectious Symptomatic Quarantined (ISQ)*: infectious symptomatic agents isolated from the rest of the population via quarantine protocol. By isolating, these agents contribute to the reduction of the force of infection against susceptible agents.

- *Recovered (R)*: the host is no longer able to infect other individuals and is no longer susceptible to be infected by the pathogen.

Parameter	Description
β	*Contact Rate*: transmission probability per contact established.
r	*Quarantine attenuation factor*: ratio affecting the contact rate when the host is under quarantine. We'll use of factor of 0.2.
lp	*Latency Period*: time between contagion and infectiousness.
ip	*Infectious Period*: time span infectious onset and recovery.
ar	*Asymptomatic Ratio*: fraction of the infected population that show no symptoms of the disease.
trq	*Rapid Testing Execution Ratio*: fraction of the population taking the rapid test.
trsS	*Slow Testing Ratio Symptomatic*: fraction of the infected symptomatic hosts taking the PCR test.
trsA	*Slow Testing Ratio Asymptomatic*: fraction of the infected asymptomatic hosts taking the PCR test.
qr	*Quarantine Compliance Ratio*: fraction of the hosts that will comply with the quarantine mandate.
fps	*False Positive Ratio RT-PCR test*: ratio of tests turning a false positive result when the test should have been negative
fns	*False Negative Ratio of the RT-PCR test*:ratio of tests turning a false negative result when the agent is indeed carrying the virus.
sensitivity	*Sensitivity* of the rapid test compared to the RT-PCR Test.
speficifity	*Specificity* of the rapid test compared to the RT-PCR Test.
t_{Max}	*Maximum time frame*: maximum number of time steps to run the model.
g	*Contact Network*: graph representing the contacts between agents. Each vertex represents an agent, or host. Each edge is represents the contacts between agents.
timeSlowTestWait	*RT-PCR Wait Time*: average turnaround time to receive the lab test results for RT-PCR tests.
time2SlowTest	*Delta Time between fast and low tests*: time elapsed between the execution of the rapid test and the execution of RT-PCR test.
fti	*Fast Test Interval*: Number of days between fast tests.

Table 1: List of parameters describing the ABM model used for rapid testing.

Parameters

Our model can be analysed and its viability appraised by the use of parameters, variables that determine the rates of movement of agents between the different compartments in the model (Table 1).

4   Algorithm

The compartmental model described above can be translated into a set of rules and procedures in a computer language. In our case, we decided to implement the algorithm in the Wolfram Language (https://www.wolfram.com/language/). Please see the supplemental material for the actual code utilized.

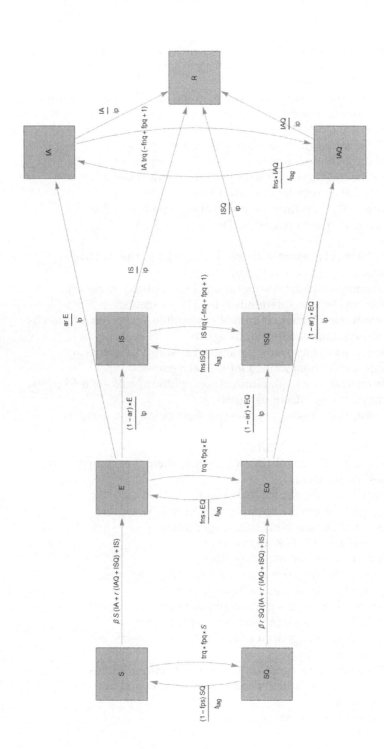

Figure 2: Compartments and population flow diagram used to design the agent based model. The texts in the edges of the graph describe the key parameters and the compartments that are involved in the calculation of the agent flow across the compartment model.

Algorithm 1: ABM Algorithm

Result: Matrix of $N = 1000$ columns, each node symbolizing an
  agent and $t_{Max} = 360$ rows portraying the state of each
  agent at time t.

Generate a Watts-Strogratz random graph of 1000 nodes, rewiring
  probability 0.04 and 18 degrees;

Determine the neighbors of each node;

Set up the initial status of each node to susceptible;

Set up three random nodes as infected (asymptomatic);

Determine the latency period (lp) of each node based on a
  log-normal distribution of 3 days and $\sigma = 1.35$;

Determine the infectious period (i) of each node based on a
  log-normal distribution of 10 days and $\sigma = 1.35$;

while $t <= t_{Max}$ **do**

> /* Identify agents to be infected by the pathogen   */
> Identify agents in each compartment;
> Determine which non-quarantined susceptible agents are
>   infected by non-quarantined infectious agents;
> Determine which quarantined susceptible agents are infected by
>   non-quarantined infectious agents;
> Determine which non-quarantined susceptible agents are
>   infected by quarantined infectious agents;
> Determine which quarantined susceptible agents are infected by
>   quarantined infectious agents;
> /* Perform tests, recovered agents do not need to
>   test   */
> /* Perform Fast Tests   */
> Determine the agents in each compartment that comply with
>   testing protocol;
> Identify test results for each agent;
> Assign false negative results randomly as per rates;
> Assign false positive results as per rates;
> /* Perform RT-PCR (Slow Tests)   */
> Determine agents requiring slow tests;
> Assign false negative results randomly as per rates;
> Assign false positive results as per rates;
> Determine timestamp for results of test;
> /* Transfer agents across compartments   */
> Move exposed agents into corresponding infectious
>   compartments;
> Move infectious agents into recovered compartments;
> Determine agents that need to be moved in and out of
>   quarantine;
> Update metric records;
> $t+ = 1$;

end

Code initialization

In this section of the code we proceed to create a matrix in which each column will represent the status of each agent, while each row, will represent the status of each agent of the network at a given timestamp. Each state (compartment that the agent belongs to) is represented by an integer. We generate a random small world graph to represent the contact network of 1000 agents. In order to optimize the code execution time we proceed to calculate up front several objects that can then be retrieve from memory during the main code block:

- We determine the list of agents that are in contact with each specific agent (neighbors).

- Agents are set as susceptible outside quarantine (compartment S).

- Three agents at random are moved to the infected asymptomatic compartment (IA). This will provide the seed for contagion across the network.

- We calculate the latency period (time that each node will stay in the Exposed compartments). Value is pulled from a log-normal Distribution with $\mu = 3$ days and $\sigma = 1.35$. Selected values are based on estimations found in the literature [7].

- Similar calculation is performed for the infectious period using a log-normal distribution with $\mu = 10$ days and $\sigma = 1.35$ [7].

Main Block

The main block of the program consists of a loop that will be executed from $t = 0$ to $t = t_{Max} - 1$. Figure 2 details the calculation used to move agents across compartments.

- Identify agents to be infected by the pathogen.

  - Determine which non-quarantined susceptible agents get infected by the pathogen. Non susceptible agents can be compromised by quarantined and non-quarantined agents. The probability of getting infected by each contact with a non-quarantined infected agent is represented by a Bernoulli distribution where the probability p determined by:

$$p = \beta$$

  - The probability of a non-quarantined susceptible agent to get infected by quarantined infected agent is affected by quarantine infection factor r, this is the same probability of a quarantined susceptible agent to be infected by a non-quarantined infected agent.

$$p = \beta r$$

- In the case that both the susceptible and the infected agent are quarantined, then the probability of contagion:

$$p = \beta r^2$$

- Move agents based on results of rapid testing. Agents that are not in quarantine status will be taking a rapid test based every fti days. Not all agents will comply with the rapid testing protocol. A specific agent will perform the rapid testing with probability trq.

  - Susceptible and Exposed agents (S / E) would be moved to quarantine if the test provides a false positive result with probability fpq.

  - Infected Agents that are not in quarantine (IA / IS) will be those where the test performs as expected, probability $p = 1 - fnq + fpq$.

- Agents that are quarantined due to the rapid test positive result will perform a slow test (RT-PCR) to confirm if they are indeed infected. The algorithm keeps track when the quarantined agents are due to get a slow test based on parameter $timeSlowTestWait = 4$.

  - Susceptible and Exposed agents in quarantine (SQ / EQ) will only remained quarantined if the RT-PCR test result is positive due to a false positive with a probability fps.

  - Infected agents in quarantine (IAQ / ISQ) will be removed from quarantine if a false negative result comes from the RT-PCR test with probability fns.

- Transfer agents across compartments.

  - Move exposed agents into the corresponding infected compartments. Agents will be moved if the latency period lp assigned during initialization has been reached. The probability of being asymptomatic is represented by parameter ar.

  - Move infected agents into the recovered compartment based on the corresponding infectious period ip assigned during initialization has been reached.

- Metrics corresponding to time t are calculated and stored in a data frame.

Parameter	Min	Max	Rationale
Contact Rate (β)	0.02	0.1	Current estimate of the Reproductive Rate $\mathcal{R}_0$ of COVID-19 is around 2.2. Given our estimation of an infectious period of 10 days and the regular degree of the network is 18, the min-max values put the observed $\mathcal{R}_0$ within the interval of interest.
Rapid Test Compliance (trq)	0.6	0.8	We expect that not everyone will comply with the testing protocol. This range is our best estimate.
Quarantine Compliance (qr)	0.6	0.9	There are some surveys asking individual if they would comply with quarantine mandates. The figure is around 0.8 for the UK.
Sensitivity of Rapid Test	0.771	0.971	Product specifications [8].
Specificity of Rapid Test	0.805	0.955	Product specifications [8].
Testing Interval (days)	1	7	We want to explore the frequency range between daily and weekly.

Table 2: Parameters used on the simulations.

Execution

Each batch of simulations is composed of 16 runs, the parameters used on the simulation and the compartmental temporal data is saved for analysis. Each batch was created by exploring the range of parameters listed in Table 2.

We can review the output of a simulation batch run to gain a better understanding on data obtained. Figure 3 presents a snapshot of the state of each agent during the each time tick of the simulation. Tallies can be made of each compartment along the temporal dimension. For our analysis, we are interested in the fraction of the population that remains susceptible at time t_{Max}. We also keep track of the fraction of the susceptible population that remains in quarantine. The fraction of susceptibles at the end of the simulation run is a measure of the effectiveness of the strategy, given the parameters under which the simulation has been performed. The proportion of the susceptible population forced into isolation by quarantine will measure the negative impact that the false positive ratio of testing will inject into the strategy.

5 Discussion and conclusions

For each batch of parameter configurations, we collected the mean and standard deviation of the final fraction of susceptible at the end of the run. A high fraction of susceptible and quarantined susceptible at the end of the simulation run indicates that the strategy has been successful to contain the spread of the virus through the contact network. The fraction of the quarantined susceptible will provide a measure of negative impact to the reduced activity on the network due to isolation of these agents. We proceeded to analyze how this final metric is affected by the changes on key parameters.

⟨| S → ■, SQ → ■, E → □, EQ → ▨, IA → ▨, IAQ → ■, IS → ▨, ISQ → ■, R → ■ |⟩

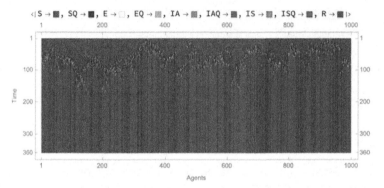

(a) This matrix plot tracks each agent's status (column) across the simulation run. The colors represent the compartment where the agent is located. Susceptibles (S), susceptibles quarantined (SQ), exposed (E), exposed quarantined (EQ), infected asymptomatic (IA), infected asymptomatic quarantined (IAQ), infected symptomatic (IS), infected symptomatic quarantined (ISQ), recovered (R).

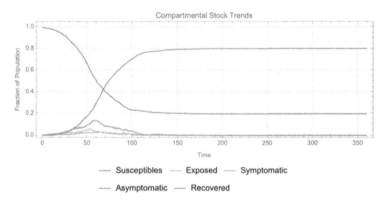

(b) Agents on each compartment can be tallied at each time snapshot and aggregated into time series. In the chart above, we can see the fraction of the population under each compartment group.

Figure 3: Example of results for a simulation run.

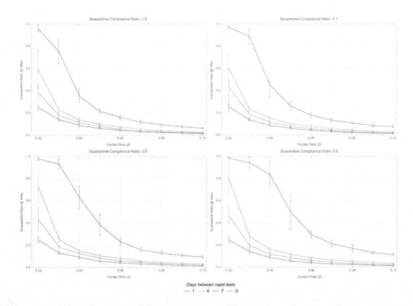

Figure 4: Impact on the remaining fraction of susceptibles on changes in the values of the frequency of testing, quarantine compliance and contact rates. 1 represents daily testing, 7 weekly, 0 represents only RT-PCR tests. Analysis performed for rapid test execution ratio = 0.8 and rapid test sensitivity = 0.971 and specificity = 0.955.

Frequency of rapid testing and quarantine compliance. Figure 4 shows the how the fraction of susceptible at the end of the simulation run is affected by not only the frequency of testing, but also by the quarantine compliance ratio and the contact rate β. Daily testing is key to the success of this strategy, but as the contact rate of the pathogen increases, this strategy quickly ceases to be of use. Recall that the reproductive rate of a disease is proportional to the contact rate and inversely proportional to the infectious period [9]. We can observe that the quarantine compliance ratio helps extend efficacy of the rapid testing. Beyond a certain value of the contact rate, a complete different strategy would be needed.

Rapid testing Compliance. Figure 5 shows the sensitivity of the strategy to changes of the fraction of the population willing to perform daily testing. This factor increases its importance as the contact rate of the pathogen increases in value, but beyond a certain threshold of the infectiousness of the disease, higher levels of compliance will no longer have an effect in making the strategy more successful.

Specificity. The rapid test specificity measures the ability of the test to generate a negative result for those individuals that are not infected. In other words, a 90% specificity will return 10% of false positives for those agents that are not infected by the disease and should have received a neg-

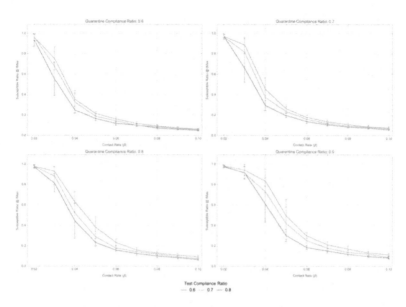

Figure 5: Sensitivity analysis on the remaining fraction of susceptibles based due to changes on rapid testing execution ratio, quarantine compliance and contact rates. Analysis performed for daily testing frequency, rapid test sensitivity = 0.971 and specificity = 0.955.

ative test back. Figure 6 indirectly shows the value of reducing the force of infection in the population. A higher level of false positives puts more agents in quarantine, affecting the contact rate as a minimum by the quarantine attenuation factor. This indicates that for more contagious pathogens, other types of strategies would need to be explored. For example, complete lockdown of a hot spot region with a cessation of activities for two infectious cycles and complete testing of the population would be able to cull the epidemic. This type of drastic strategy might not be acceptable to the population, but could be the only available tool that a society can use in the case of a highly contagious disease.

Sensitivity. The rapid test sensitivity affects the fraction of false negatives that the test generates. A test with 90% sensitivity will result on 10% of the tests providing a false negative result. This means that 10% of the agents that should have received a positive result back will get a negative test result. It is an interesting observation that the results observed do not have a high impact on the success of the strategy. The daily frequency of testing could explain the reduced effect of in the success of the strategy for a test with a lower level of sensitivity, due to the assumption of independence of a false negative test on an individual from prior tests performed, then the probability of n consecutive false negative tests would be $(1 - sensitivity)^n$.

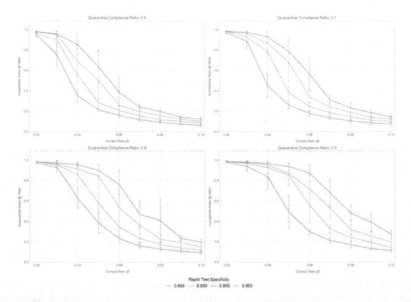

Figure 6: Impact on the remaining fraction of susceptible on changes in the values of the rapid testing specificity, quarantine compliance and contact rates. Analysis performed for daily testing frequency, rapid test execution ratio = 0.8 and specificity = 0.955.

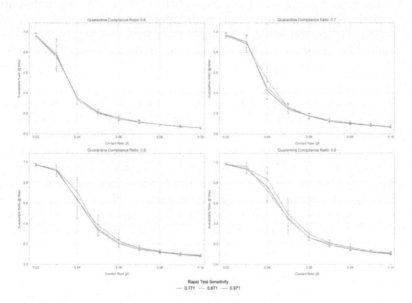

Figure 7: Impact on the remaining fraction of susceptible on changes in the values of the rapid testing sensitivity, quarantine compliance and contact rates. Analysis performed for daily testing frequency, rapid test execution ratio = 0.8 and specificity = 0.955.

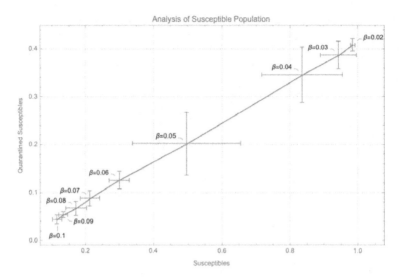

Figure 8: Relationship between the quarantined susceptible population vs. total quarantined population. Analysis performed for daily testing, rapid test execution ratio = 0.8, quarantine compliance = 0.9, specificity = 0.955, sensitivity = 0.971.

Overall impact on the mobility of the population. Figure 8 shows that approximately 41% of the susceptible population would be in quarantine if this strategy is implemented. The daily frequency of testing and the inevitability false positive results inherent to the diagnostic test causes such a high fraction of the susceptible population to have to be mistakenly placed in isolation.

We can conclude that a strategy of massive repetitive testing using rapid diagnostics has several hurdles to overcome. We have observed that a highly infectious disease would overwhelm the system, the high frequency of testing would not be sufficient to control the spread of the pathogen through the population. A false positive rate of 8.7% that we can expect from this type of tests can drive 2/5 of the susceptible population into unnecessary isolation. Quick, cheap tests still have an important role to play beyond the strategy discussed on this paper. The faster the test results are obtained, the sooner an infected person can be placed under the appropriate treatment guidelines.

6   Supplementary Materials

The Wolfram Language code can be found in https://github.com/dzviovich/COVID-RapidTest.

7 Acknowledgments

We want to thank Anton Antonov for the development of several utilities made available to the public. (https://mathematicaforprediction. wordpress.com/) and Joe Norman for his help and input on the outline and introduction of this document.

Bibliography

[1] Tognotti, Eugenia. "Lessons from the History of Quarantine, from Plague to Influenza A." Emerging Infectious Diseases, vol. 19, no. 2, Feb. 2013, pp. 254–59. DOI.org (Crossref), doi:10.3201/eid1902.120312.

[2] Norman Joseph, et al., "Systemic risk of pandemic via novel pathogens–Coronavirus: A note", New England Complex Systems Institute Jan. 2020

[3] Afzal, Adeel. "Molecular Diagnostic Technologies for COVID-19: Limitations and Challenges." Journal of Advanced Research, vol. 26, Nov. 2020, pp. 149–59. DOI.org (Crossref), doi:10.1016/j.jare.2020.08.002.

[4] Kucirka, Lauren M., et al. "Variation in False-Negative Rate of Reverse Transcriptase Polymerase Chain Reaction–Based SARS-CoV-2 Tests by Time Since Exposure." Annals of Internal Medicine, vol. 173, no. 4, Aug. 2020, pp. 262–67. DOI.org (Crossref), doi:10.7326/M20-1495.

[5] Katz, Andrew P., et al. "False-positive Reverse Transcriptase Polymerase Chain Reaction Screening for SARS-CoV-2 in the Setting of Urgent Head and Neck Surgery and Otolaryngologic Emergencies during the Pandemic: Clinical Implications." Head & Neck, vol. 42, no. 7, July 2020, pp. 1621–28. DOI.org (Crossref), doi:10.1002/hed.26317.

[6] CDC. "Discontinuation of Isolation for Persons with COVID-19 Not in Healthcare Settings".
https://www.cdc.gov/coronavirus/2019-ncov/hcp/
disposition-in-home-patients.html

[7] Lauer, Stephen A., et al. "The Incubation Period of Coronavirus Disease 2019 (COVID-19) From Publicly Reported Confirmed Cases: Estimation and Application." Annals of Internal Medicine, vol. 172, no. 9, May 2020, pp. 577–82. DOI.org (Crossref), doi:10.7326/M20-0504.

[8] Abbott Laboratories, "BibaxNow™Procedure Card", https://www.fda.gov/media/141570/download

[9] Mishra, S., et al. "The ABC of Terms Used in Mathematical Models of Infectious Diseases." Journal of Epidemiology & Community Health, vol. 65, no. 1, Jan. 2011, pp. 87–94. DOI.org (Crossref), doi:10.1136/jech.2009.097113.

An Alignment-Free Method for Classification of the SARS-CoV-2 Genome

Daniel Lichtblau*

Wolfram Research, 100 Trade Center Dr, Champaign, IL, 61820, USA.
*Corresponding author: danl@wolfram.com

We apply a recent alignment-free method of genomic comparison to sequences of SARS-CoV-2 as well as other sequences from the Coronaviridae family. We show that this method, while approximate, can enable fast and accurate classification. We illustrate how it might be applied in the search for the possible intermediary host or hosts. We also use this methodology at a finer level, to create a phylogenetic tree from SARS-CoV-2 sequences taken over a period of time and from geographically distinct locations. This can help to determine routes by which the disease has traveled and also help to chart the course of mutations both in time and geography, thus providing useful information in the realms of epidemiology and public health policy. As an important application we analyze geographical locations in which certain more infectious variants have appeared. By comparing fraction of variant appearances against date of collection we can estimate the rate at which such variants are spreading.

Keywords: Alignment-free methods, genome comparison, SARS-CoV-2, chaos game representation, phylogenetic tree, dimension reduction, multidimensional scaling.

1 Introduction

Since it was first diagnosed in early December of 2019, and sequenced in late December of that same year, the SARS-CoV-2 virus and its attendant COVID-19 pandemic has spread across the planet. Over this time period there have been innumerable studies of its various features. These include (but are by no means limited to) the physics of transmission, the biochemistry of its inner workings, epidemiology of the spread and various mitigation policies, vaccine research and protocols, therapy studies, and genomic classifications. This last is the topic of this paper. Within this subfield alone one finds again a wide variety of methods employed, including alignment-based comparison to existing coronavirus genomes, alignment-free comparisons, searches for possible reservoir hosts or common genetic ancestors, and so forth. We will utilize a particular alignment-free comparison method developed by the author in [15]. This in turn has roots in prior work such as [1, 4, 9–11, 14, 19, 21–23].

A key idea behind the method is to capture some aspects of the sequences, perhaps as images or numeric vectors, and apply image and/or signal processing methods in a way that is fast and allows for distance-based comparisons. One family of methods, seen in the above references, uses the Frequency Chaos Game Representation (FCGR) [1, 4] (based on earlier work by Jeffrey [9]). This creates images with certain fractal properties that capture frequencies of k-mers for modest values of k. A number of different processing methods have then been deployed in order to classify these images; references [1, 4, 10, 11, 21–23] how several of these and also give some indication of their variety.

As described in [15], our method starts with these FCGR images. We reduce dimension by applying a Fourier Discrete Cosine Transform (DCT) to each image matrix, retaining only low frequency components. We then flatten the resulting matrices into vectors and use the Singular Value Decomposition (SVD) to further reduce dimension. The vectors that result from this are used in clustering, both by Multidimensional Scaling (MDS) and by creating phylogenetic trees. This can be applied to inferring taxonomy information for new sequences, given a reference database for known genomes [21, 22].

The importance of the present work is that it involves fast computations, capable of working with hundreds or thousands of genomes at a time. It avoids the much slower construction of genome pair alignment distances and thus provides a set of tools that can rapidly home in on interesting clustering or other features. This in turn allows one to observe trends over time or across geographic locations that might be much more troublesome to compute at scale with alignment-based methods. Moreover these methods can provide rapid information that might be used to direct efforts that would require more time- or resource-consuming methods.

The outline of this paper is as follows. We first review the methods from [15] in order to make this work self-contained. After that we apply it to get a broad classification of SARS-CoV-2 specimens in the Coronaviridai family, showing, among other things, nearest neighbors found to date in databases. We then show how one might find progressions of variants across both time and geography. We use this approach to show proximity to recently discovered variants that have been shown to be particularly infectious. We further investigate the spread of one such variant in certain locations.

A powerful set of related methods appears in [19], with strong result shown for several tests both in species recognition and phylogeny tree construction (which the authors have made available for benchmark purposes). The tandem of FCGR and SVD is used in [22] on a set of 400 of Human Papillomavirus (HPV) genomes from 12 strains, where it attains perfect classification at the strain level (this data set was also classified with no errors in [15]).

Recent work on this topic has appeared that uses related methods. In [2,8,12,16,18,25,26] there are analyses and genomic comparisons to infer how SARS-CoV-2 is related to other coronaviruses and also to show nearest known genomic relatives. In [7] alignment-free methods are used to group variants of SARS-CoV-2 by geography. References [6,13,17] analyze the spread of recent variants P.1, B.1.1.7 and B.1.351/501Y-V2 from Brazil, the United Kingdom and South Africa, respectively.

All computations herein were performed using version 12 of Mathematica [24]. These were run on a desktop machine with a 3 GHz processor, 16 Gb RAM, running under the Linux operating system. Wolfram Language code for all experiments is available at https://notebookarchive.org/id/2021-02-5kj28s7. Sequences utilized were obtained from GenBank [3] and GISAID [5,20]. Sequence accession identifiers and related metadata are provided in the supplemental section. We thank the creators and maintainers of those sites, as well as the many laboratories around the world that have collected, sequenced and contributed the genomic data.

2 Methods and materials

The method we use converts genome sequences into numerical vectors of length 40. The details, including a complexity analysis, are provided in [15]. In brief, we convert each sequence into an image matrix using the FCGR, and reduce matrix dimension by applying the Fourier DCT to each such image with the mean subtracted. Once all genomes have been processed in this way, we flatten the reduced matrices into vectors and apply the SVD to further reduce dimension. We now have vectors of length 40. We use these in two ways. One is to reduce to two or three dimensions by MDS, as this allows us to visualize a good estimate of relative proximities. The other creates phylogenetic trees as dendrograms based on vector distances. We elaborate on these steps below.

Frequency Chaos Game Representation

Given a genetic sequence, a method to convert to an image was presented by Joel Jeffrey in the late 80's [9]. Label the corners of a square with the four nucleotide bases. One starts in the middle of the square. Reading the sequence, we mark a dot midway from current position to the corner labeled with the next base. A modest refinement, used herein, in effect discretizes the CGR and is called the Frequency Chaos Game Representation (FCGR) [1,4]. Here one "pixelates" with a square of side-length equal to a power of two. We illustrate in Figure 1 using a SARS-CoV-2 genome sequence and three other coronavirus sequences, using sides of length 2^7 (that is, a pixelation level of 7). This has the effect of creating a fractal-like

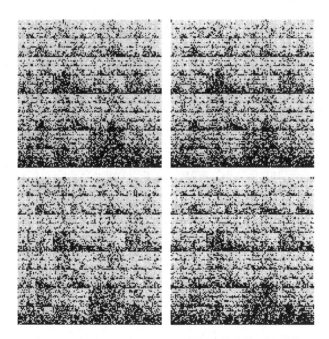

Figure 1: FCGR images clockwise from upper left: SARS-CoV-2 reference sequence (Wuhan), RatG13, Pangolin isolate MP789, SARS-CoV BJ182b

pattern since each dot represents a nucleotide sequence of length 7, and subsquares capture suffix subsequences from their parent square.

CGR images have been studied in a number of ways and have some interesting properties. FCGR images have been found to be particularly useful in genomics. This derives from the empirical fact that similar genomes tend to show similar fractal-like patterns (up to the degree of resolution from the pixelation).

Discrete Cosine Transform

Dimension reduction using the Fourier DCT will give rise to coarser images. In effect, higher frequency detail is removed from the images, leaving the coarser main frequencies. We illustrate in Figure 2 (for purposes of visual comparison with the original images we transform back using the inverse DCT).

Singular Values Decomposition

After reducing dimension by Fourier DCT, we further reduce using the SVD. This step requires the full set of sequences on which we work, so we use here the 72 sequences from experiment 1 (to be described below). The result will be vectors of 40 elements, and from that we can reconstruct the

Figure 2: Images from Fig. 1 with dimensions reduced by Fourier DCT

reduced dimension images. We show this in Figure 3, again using the same four sequences.

What we have after these steps is a lossy, but nonetheless useful, representation of a set of genomic sequences. They act as a set of "signatures", with the nice property that genomic relatives can be discerned as numeric vector relatives using any suitable distance function; for our purposes, the usual Euclidean distance does quite well [15]. Because they are relatively short they can be used for purposes of look-up via kD trees (see [15] for further details). Our use below will be to gauge relative proximities of genome pairs and clusters.

Multidimensional Scaling

We use these signature vectors below in two important ways. One is to derive three dimensional visualizations of genome "proximities". For this purpose we apply a further dimension reduction using MDS. It too is based on SVD. In effect it projects from higher dimensions in a way that optimally preserves pairwise distances. The resulting picture is useful for noticing trends, particularly if the dimension-reduced points are colored in a way that represents temporal or geographical relationships. In our uses the colors will be based on time, specifically, the dates in which the genome samples were obtained.

Figure 3: Images from Fig. 2 with dimensions further reduced by SVD

Phylogenetic Trees

Our other use of these signature vector sets is for constructing phylogenetic trees. This is accomplished by looking for closest neighbors in terms of distance (again, we use Euclidean, but others such as cosine work well). Each pair is made into a branch, with edge length scaled to be commensurate with the distance. The pair is replaced by its center, and the process continues. In the tree that results, all branches that are not terminal split into a pair of subtrees. The resulting trees group close sequences as neighbors, and relative genomic distance for a given pair can be inferred by lengths of branching from the nearest common split. Our layout of these trees will be from left to right, that is, tree root on the left, increased branching as we go left-to-right, and sequences on the right.

Code

All code used in the experiments below is available in a Mathematica notebook from https://notebookarchive.org/id/2021-02-5kj28s7. The code is fairly straightforward. Technical functions used are FourierDCT and SingularValuesDecomposition. It also makes use of several functions found in the Wolfram Function Repository. They are ImportFASTA, PhylogeneticTreePlot, FCGRImage and MultidimensionalScaling. It also uses data downloaded from the GISAID web site. The experiments and supplemental information indicate when the downloads were done and what sequences (by accession ID) were used.

Data

Our data is taken from the NCBI resource GenBank [3] and from the GISAID web site [5, 20]. We downloaded many sequences from each, and further manipulated to color code by date, geography, or for other distinguishing features. Data from GenBank is publicly accessible via the Wolfram Data Repository, and code for this purpose may be found in the notebook. All accession IDs are provided in the supplemental section.

3   Experiments

Near Relatives of SARS-CoV-2

Once the SARS-CoV-2 genome was first sequenced (before it even had this name), a natural topic to investigate was whether it was either already a known disease or, if not, where it fit into the spectrum of relatives. Several studies have appeared on this topic, including [2, 12, 16, 18, 25, 26] and the very recent preprint [8]. We use many of the sequences that appear in these studies as well as several others, in order to deduce approximate relatives. While some of these go further in terms of localized comparisons and analysis of genome differences, they involve more costly methods. Once the sequences have been obtained from GenBank and GISAID, our code takes but a few seconds to produce the phylogenetic tree of Figure 4. The colors are as follows. Early SARS-CoV-2 genomes are black. Genomes that have been claimed as close relatives are red. MERS and related are green. SARS genomes are blue. Bat SARS and related are purple.

The phylogenetic tree in Figure 4 is quite similar to others that have been published, and in particular has similar placing as in [8] of the closest known relatives, represented by RaTG13, the recently sequenced RshSTT200, and prior known relatives from bat and pangolin specimens. We see moreover that closer relatives are grouped together and, in particular, the SARS-CoV-2 genomes are far closer to one another than to any others.

Charting the Spread of Variants

Once the virus has had many generations to replicate and spread in geography, it is expected that mutations will accrue, and that they will appear at similar times, both in geographically proximate locations, with travel patterns and dates indicating directions of spread. We show an example where geographic isolation has been enforced, in Australia. We have 350 sequences downloaded from GenBank, dated between 2020-02-21 and 2020-

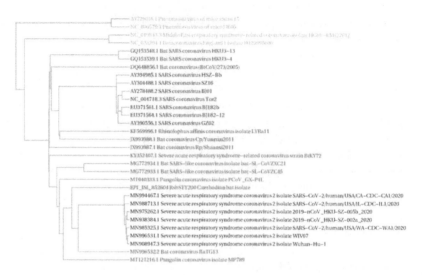

Figure 4: Phylogenetic tree of SARS-CoV-2 and some coronaviridae relatives

10-06. We apply Multidimensional Scaling to reduce to three dimensions and plot the locations colored by date of collection in Figure 5.

In order to understand the directions of change over time, the earlier dates are colored reddish-brown, and they proceed to yellowish-brown, then green, then cyan, and finally blue. From this color scheme we can see that the first collected genomes were related, as they mostly appear near one another near the upper left. Moreover the virus has mutated over time and split into what appear to be distinct branches. Australia has stopped most international entry since March of 2020 so it is likely that these mutations largely arose within the country rather than from an influx due to travel from other countries.

The B.1.1.7 invasion

In late November of 2020 reports emerged from the United Kingdom to the effect that a new variant seemed to be more contagious than prior ones. Further sequencing in December indicated that it was rapidly becoming the dominant form of the virus in parts of England. Originally called VOI202012/01 (for "variant of interest"), it is now known as the B.1.1.7 lineage, and has now spread to several dozen countries around the world [17]. We obtained from the GISAID site 129 sample sequences of this lineage, all collected in various parts of England during January 2021. Below we show two case studies comparing sequenced SARS-CoV-2 genomes from Ireland and Florida to these reference B.1.1.7 sequences.

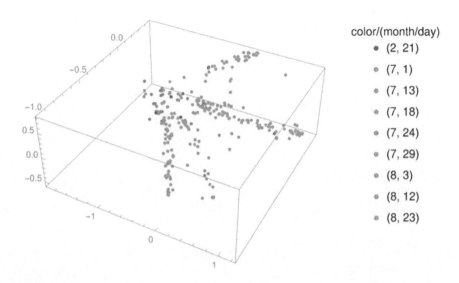

color/(month/day)
- (2, 21)
- (7, 1)
- (7, 13)
- (7, 18)
- (7, 24)
- (7, 29)
- (8, 3)
- (8, 12)
- (8, 23)

Figure 5: Spread of variants in Australia (brown-red-yellow are older than green-cyan-blue)

The B.1.1.7 Variant in Ireland

We downloaded all sequences from Ireland listed both as complete and high coverage and placed on the GISAID site as of 2021-02-01, with collection dates on or after 2020-12-01. There were in total 369 such. We color by date, with brown-yellow-red for older collection dates and green for more recent. We color the B.1.1.7 sequences as gray dots. The 3-dimensional MDS plot in Figure 6 indicates that several recently collected genome specimens from Ireland are very likely in the B.1.1.7 lineage.

The spread of gray dots indicate that the B.1.1.7 lineage is itself quite variable. This spread is also seen in the phylogenetic tree plot of Figure 7; here we decimate the collection by a factor of eight to make it easier to read. Again, we see that a number of sequences from Ireland fall well within the B.1.1.7 sequences. If we restrict attention to sequences collected after 2021-01-01, then more than half appear to fall among the B.1.1.7 lineage. Moreover, no sequence from prior to 2020-12-22 appears in this part of the tree. So it is clear that the percentage of genomes falling into the B.1.1.7 lineage is increasing. As the most recent collection date is 2021-01-22 and the date of download is 2021-02-01, it is almost certainly the case that the percentage of cases in Ireland from this lineage has since increased further.

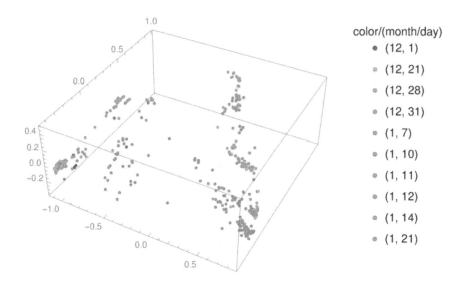

Figure 6: MDS view of genomes from Ireland vs B.1.1.7 genomes from England (gray)

It should be noted that, while the method we have used involves lossy compression, it can be validated. One can, for example, use string edit distances to show that the sequences in the tree that appear to be of the B.1.1.17 lineage actually are. Several neighboring pairs were checked, and in all cases the string edit distances were in the zero-to-five range. This is, needless to say, a far slower computation than those we have illustrated above.

The B.1.1.7 variant in Florida

We downloaded all sequences listed both as complete and high coverage, collected in the state of Florida on or after 2020-12-01 and placed on the GI-SAID site as of 2021-02-01. There were in total 258 such. The 3-dimensional MDS plot in Figure 8 indicates that several recently collected genome specimens from Florida are perhaps in the B.1.1.7 lineage.

A projection to three dimensions can sometimes be misleading. So again we also construct a phylogenetic tree in Figure 9; again we decimate by a factor of eight for readability. This tree also makes plausible that a number of Florida genomes come from the B.1.1.7 lineage.

Using an edit distance on the Florida sequences that appear within the B.1.1.7 lineage shows an interesting phenomenon. They tend to be

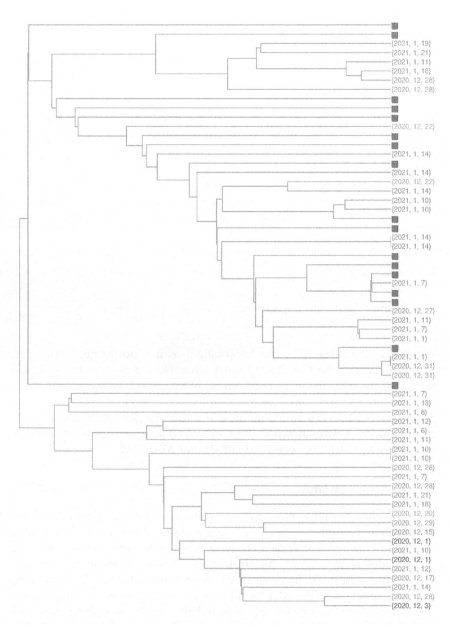

Figure 7: Phylogenetic tree of recent genome sequences from Ireland along with B.1.1.7 sequences (gray)

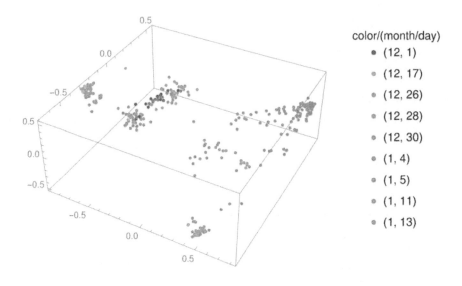

Figure 8: MDS view of genomes from Florida vs B.1.1.7 genomes (gray)

separated by a distance in the 30-50 range. This is not terribly close, but certainly closer than they are to most other Florida sequences (and also typical of the edit distances between pairs from the B.1.1.7 lineage). So it would seem likely that at least a few of these recent Florida sequences are in fact the B.1.1.7 variant. We remark also that the genome sequence data from Florida is relatively scant, with no new sequences appearing either in GISAID or GenBank between 2020-02-01 and 2020-02-08.

California and variants

California is reported to have a number of cases from two strains believed to be more virulent than the rest. They are the British B.1.1.7 and also the B.1.429 lineage that appears to have arisen in southern California. We downloaded from GISAID all complete B.1.429 sequences collected in California since 2021-01-01 and submitted no later than 2020-02-01. There were 215 such sequences. In order to get a non-overlapping set of other recent sequences from that state we downloaded all complete sequences collected in California since 2020-12-01 and placed in the GenBank (rather than GISAID) repository as of 2021-02-12; there were 932 such. The color scheme for the MDS plot in Figure 10 uses blue for the GISAID B.1.429 variant, gray for the GISAID B.1.1.7 variant, and a color range based on dates for the GenBank sequences. The red-brown to brown-yellow to yellow-green

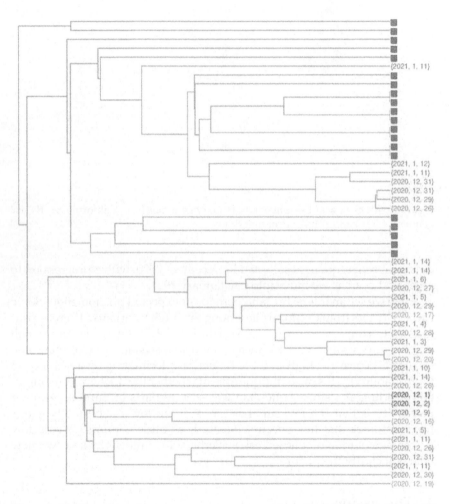

Figure 9: Phylogenetic tree recent genome sequences from Florida along with B.1.1.7 sequences (gray)

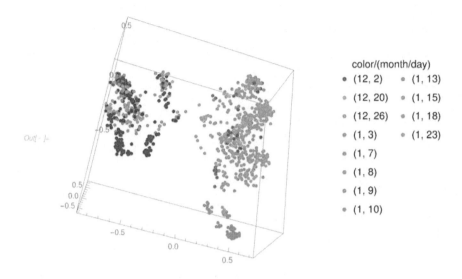

Figure 10: MDS view of December and January genomes from California vs. B.1.1.7 genomes (gray) and B.1.429 genomes (blue)

dots represent sequences collected in December 2020, with a progression to green and cyan as we move into late January 2021.

Here we see an interesting and perhaps unexpected phenomenon. Some earlier clusters from California lie among the B.1.429 variants. There appear to be quite a few later sequences also in the B.1.429 cluster, while only a small number are within the B.1.1.7 lineage. his also is seen in a phylogenetic tree (Figure 11) constructed from these sequences (decimating by a factor of 15 for readability). This raises the question of whether the B.1.1.7 variant might be spreading at a slower rate than was seen in parts of Europe. Another possibility is that the raw numbers are still too low to accurately gauge from genomes that have been sequenced, as these comprise a very low percentage of total diagnosed cases. Perhaps in another week or two new data will clarify the picture.

4   Discussion

We have applied an alignment-free method for the study and comparison of genomes to many aspects of the genomics of SARS-CoV-2. We first showed how a fast phylogenetic tree construction can separate members of the coronaviridae family, even beyond the subgenus sarbecovirus level. We used similar methods to visualize the mutation spread of SARS-CoV-2 in Australia. We then focused on three locations, Ireland, Florida, and California, that have shown signs of a particular strain (B.1.1.7) becoming dominant

Figure 11: Phylogenetic tree recent genome sequences from California along with B.1.1.7 sequences (gray) and B.1.429 sequences (blue)

(with competition from the B.1.419 strain in California). In locales such as Ireland, where substantial genome collection and sequencing are done, we are able to obtain quite useful information about the relative spread rates (corresponding to levels of contagiousness) of the new vs. previously present variants of SARS-CoV-2. This can be valuable information e.g. for informing public policy in terms of medical preparation, immunization, and the like. Even where raw data is more scarce we are still able to see rough trends, as appear in the studies of Florida and California.

While not in general as accurate as alignment-based methods, we still obtain useful information fast and with relatively low computational resources. This capability can be quite important for informing and directing further study using more expensive resources. It can moreover be applied in situations where genome sequences might be quite long, e.g. for classifying newly discovered bacterial sequences. The applications we presented include an initial classification of the SARS-CoV-2 novel coronavirus within the Coronaviridae family, visualizing mutation distances, and allowing to estimate the rate at which more contagious variants become dominant in given geographical locales. All of these are important in terms of understanding the viral genome and its spread. The last of the applications can even be used to estimate relative levels of contagiousness.

It should be mentioned that these methods do come with weaknesses. As with any lossy compression, information is lost that, in some cases, might be important. There is also an issue of genome sequence quality. Experiments in [15] indicate that partial segments, and "full" genomes with unknown nucleotides (denoted by N rather than the usual A,C,G,T/U) can be quite useful for determining genus and species. But when we get to the level of variants of a sarbecovirus, imperfect sequences can skew results using our methodology (in particular MDS might show pairs as being closer than actual genetic distance would warrant). Nonetheless this approach provides useful tools for fast analyses that may in turn suggest areas to pursue with slower alignment-based methods.

Acknowledgments

I thank my colleague M. Bahrami for encouraging me to write up this research. Earlier (and cruder) analyses, using similar methods, were presented in posts to the Wolfram Community forum in February and April of 2020.

Bibliography

[1] Jonas S. Almeida, João André Carriço, António Maretzek, Peter A. Noble, and Madilyn Fletcher. Analysis of genomic sequences by Chaos Game Representation. *Bioinformatics*, 17(5):429–437, 2001.

[2] K. G. Andersen, A. Rambaut, W. I. Lipkin, E. C. Holmes, and R. F. Garry. The proximal origin of SARS-CoV-2. *Nature Medicine*, 26:450–452, 2020.

[3] Dennis A. Benson, Mark Cavanaugh, Karen Clark, Ilene Karsch-Mizrachi, David J. Lipman, James Ostell, and Eric W. Sayers. GenBank. *Nucleic Acids Research*, 45(D1):D37–D42, 2016.

[4] Patrick J. Deschavanne, Alain Giron, Joseph Vilain, Guillaume Fagot, and Bernard Fertil. Genomic signature: characterization and classification of species assessed by Chaos Game Representation of sequences. *Molecular Biology and Evolution*, 16(10):1391–1399, 1999.

[5] S. Elbe and G. Buckland-Merrett. Data, disease and diplomacy: GISAID's innovative contribution to global health. *Global Challenges*, 1:33–46, 2017.

[6] N. R. Faria, I. Morales Claro, D. Candido, L. A. Moyses Franco, P. S. Andrade, T. M. Coletti, C. A. M. Silva, F. C. Sales, E. R. Manuli, R. S. Aguiar, N. Gaburo, C. da C. Camilo, N. A. Fraiji, M. A. Esashika Crispim, M. d P. S. S. Carvalho, A. Rambaut, N. Loman, O. G. Pybus, and E. C. Sabino. Genomic characterisation of an emergent SARS-CoV-2 lineage in manaus: preliminary findings. *virological.org*, 2021.

[7] G. Hahn, S. Lee, S. T. Weiss, and C. Lange. Unsupervised cluster analysis of sars-cov-2 genomes reflects its geographic progression and identifies distinct genetic subgroups of SARS-CoV-2 virus. *Genetic Epidemiology*, 1:1–8, 2021.

[8] V. Hul, D. Delaune, E. A Karlsson, A. Hassanin, P. Ou Tey, A. Baidaliuk, F. Gámbaro, V. Tan Tu, L. Keatts, J. Mazet, C. Johnson, P. Buchy, P. Dussart, T. Goldstein, E. Simon-Lorière, and V. Duong. A novel SARS-CoV-2 related coronavirus in bats from cambodia. *bioRxiv*, page 18, 2021.

[9] H. Joel Jeffrey. Chaos game representation of gene structure. *Nucleic Acids Research*, 18(8):2163–2170, 1990.

[10] Jijoy Joseph and Roschen Sasikumar. Chaos game representation for comparison of whole genomes. *BMC Bioinformatics*, 7:243, 2006.

[11] Rallis Karamichalis, Lila Kari, Stavros Konstantinidis, and Steffen Kopecki. An investigation into inter- and intragenomic variations of graphic genomic signatures. *BMC Bioinformatics*, 16:246, 2015.

[12] S. Kumar, Q. Tao, S. Weaver, M. Sanderford, M. A. Caraballo-Ortiz, S. Sharma1 S. L. K.Pond, and S. Miura. An evolutionary portrait of the progenitor SARS-CoV-2 and its dominant offshoots in COVID-19 pandemic. *bioRxiv*, page 20, 2020.

[13] B. Larsen and M. Worobey. Phylogenetic evidence that B.1.1.7 has been circulating in the United States since early- to mid-November. *virological.org*, 2021.

[14] D. Lichtblau. Linking Fourier and PCA methods for image look-up. In *2016 18th International Symposium on Symbolic and Numeric Algorithms for Scientific Computing (SYNASC)*, pages 105–110. IEEE, 2016.

[15] D. Lichtblau. Alignment-free genomic sequence comparison using FCGR and signal processing. *BMC Bioinformatics*, 20:742, 2019.

[16] R. Lu, X. Zhao, J. Li, P. Niu, B. Yang, H. Wu, W. Wang, H. Song, B. Huang, N. Zhu, Y. Bi, X. Ma, F. Zhan, L. Wang, T. Hu, H. Zhou, Z. Hu, W. Zhou, L. Zhao, J. Chen, Y. Meng, J. Wang, Y. Lin, J. Yuan, Z. Xie, J. Ma, W. J. Liu, D. Wang, W. Xu, E. C. Holmes, PG. F. Gao, G. Wu, W. Chen, W. Shi, and W. Tan. Genomic characterisation and epidemiology of 2019 novel coronavirus: implications for virus origins and receptor binding. *The Lancet*, 395:565–574, 2020.

[17] A. O'Toole, M. Kramer, and et al. Tracking the international spread of sars-cov-2 lineages B.1.1.7 and B.1.351/501Y-V2. *virological.org*, 2021.

[18] G. S. Randhawa, M. P. M. Soltysiak, H. El Roz, C. P. E. de Souza, K. A. Hill, and L. Kari. Machine learning using intrinsic genomic signatures for rapid classification of novel pathogens: COVID-19 case study. *PLoS ONE*, 15(4):e0232391, 2020.

[19] Gurjit S. Randhawa, Kathleen A. Hill, and Lila Kari. ML-DSP: machine learning with digital signal processing for ultrafast, accurate, and scalable genome classification at all taxonomic levels. *BMC Genomics*, 20(1):267, 2019.

[20] Y. Shu and J. McCauley. GISAID: Global initiative on sharing all influenza data– from vision to reality. *EuroSurveillance*, 22(13), 2017.

[21] Martin T. Swain. Fast comparison of microbial genomes using the Chaos Games Representation for metagenomic applications. In *Proceedings of the ICCS 2013*, volume 18, pages 1372–1381, 2013.

[22] Watcharaporn Tanchotsrinon, Chidchanok Lursinsap, and Yong Poovorawan. A high performance prediction of HPV genotypes by Chaos Game Representation and singular value decomposition. *BMC Bioinformatics*, 16:71, 2015.

[23] Yingwei Wang, Kathleen Hill, Shiva Singh, and Lila Kari. The spectrum of genomic signatures: from dinucleotides to Chaos Game Representation. *Gene*, 346:173–185, 2005.

[24] Wolfram Research. Mathematica 12.2, 2021.

[25] T. Zhang, Q. Wu, and Z. Zhang. Probable pangolin origin of SARS-CoV-2 associated with the COVID-19 outbreak. *Current Biology*, 30:1346–1351, 2020.

[26] H. Zhou, X. Chen, T. Hu, J. Li, H. Song, Y. Liu, P. Wang, D. Liu, J. Yang, E. C. Holmes, A. C. Hughes, Y. Bi, and W. Shi. A novel bat coronavirus reveals natural insertions at the S1/S2 cleavage site of the spike protein and a possible recombinant origin of HCoV-19. *bioRxiv*, page 16, 2020.

The Role of Mobility Patterns in the Formation of Bubbles During the COVID-19 Pandemic

Olha Buchel*, Leila Hedayatifar and Yaneer Bar-Yam

New England Complex Systems Institute, 277 Broadway St, Cambridge, MA, USA
http://www.necsi.edu/
*Corresponding author: olha@necsi.edu

Multiscale quarantine social bubbles became popular mitigation methods during the COVID-19 pandemic. At the micro level, people were asked to form their own small social bubbles to minimize infections rates locally. At the macro level, states and countries formed travel zones to minimize the transmission between remote locations. The formation of these bubbles was often based only on infection rates, disregarding common mobility patterns. In this chapter we demonstrate that the infection dynamics at different locations follow divergent growth patterns, largely due to highly-varying mobility patterns. We examine heterogeneity of movement and disease diffusion and propose a multi-level quarantine strategy which accounts for human mobility patterns and the severity of COVID-19 contagion in different areas. Specifically, we analyze the dynamics of mobility patterns during the COVID-19 outbreak by applying the Louvain method with modularity optimization to the weekly mobility networks obtained from SafeGraph data. The analysis of mobility patterns helped us identify natural boundaries of human interactions during the pandemic and observe what effect quarantine policies and lockdowns have on mobility patterns and disease diffusion. Using the locations of confirmed cases, we also identify natural boundaries of high risk and zero-COVID bubbles. Identification of bubbles at multiple scales and high and low risk areas provides policy makers with valuable information on how to optimize travel restrictions and quarantine policies that are minimally disruptive for social and economic activities.

Keywords: Social bubbles, COVID-19 contagion.

1   Introduction

The global spread of the 2019 novel coronavirus (COVID-19) posed many challenges for public health professionals and policy makers in the design of adequate intervention methods that would limit the spread of the virus. Many of the intervention methods were targeted to restrict human mobility, especially social distancing, shelter-in-place recommendations, travel restrictions, lockdowns, border control, contact tracing, surveillance, and so on. While some of these methods have proven to be effective, they are associated with high economic and social costs and cannot be imposed indefinitely. While lockdowns are the most effective interventions, after months of lockdowns in the spring of 2020, many countries, states, and provinces turned to quarantine bubbles which were formed at multiple levels of societies: ranging from pandemic pods or quaranteams at the level of personal interrelationships to bubbles within countries and between countries [33]. When done carefully, the research shows that quarantine bubbles can effectively limit the risk of contracting SARS-CoV-2 while allowing people to have social interactions [6,32,41]. Bubbles were not limited to tourism; their goal was also to facilitate business recoveries across multiple sectors [45]. Initially, bubbles helped their citizens to balance travel risks of the pandemic with the emotional, mental, and social needs of life. But later, when the second wave of the pandemic began, COVID-19 cases started spreading across various bubbles again, exceeding the rates that were observed earlier in the spring.

The United States did not adopt a systematic strategy to boundaries for travel to reduce transmission. There were temporary restrictions places for travel that worked well in some cases such as Vermont and poorly in others such as restrictions on travelers between New York and other nearby states. In most cases, this was an ad-hoc process. While some states coordinated their efforts and acted together, others lifted restrictions independently [3,17] without analyzing positive tests, locations of current patients, and not taking advantage of the mobility patterns. Moreover, data collected, analyzed or reported about the pandemic often does not provide relevant data for determining how to establish boundaries between highly infected areas and areas with low infection rates. Quarantine policies and data related to the COVID-19 outbreak are based on state or county boundary lines as evidenced by visualizations in numerous dash- boards (Johns Hopkins University's COVID-19 dashboard [1], New York Times Coronavirus World Map: Tracking the Global Outbreak dashboard [2]).

Travel restrictions may be adopted across state, county or municipal boundaries, as has been done in Australia, Ireland and Portugal, or even within cities as has been done inside Melbourne, Australia. Often these boundary restrictions allow for essential commuters including those who need to go to work and have been successful nevertheless. Still, we can consider how to optimize such boundaries by considering the mobility patterns in the formation of social bubbles. In this project, we consider optimal social bubbles to be equivalent to natural segmentation patterns in mobility data, which are also known as functional communities in economics. Social bubbles are not new: we all used to live in the bubbles long before the pandemic. These bubbles were affected by new restrictive policies on mobility

---

[1] https://coronavirus.jhu.edu/us-map
[2] https://www.nytimes.com/interactive/2020/world/coronavirus-maps.html

and travel. When people formed social bubbles, they generally formed them from already existing relationships, relationships that make up the segmentation patterns. Previous research on social fragmentation and human mobility has already revealed geo-located communities [4,35,46] that exist at multiple scales [13,20]. People in these segments (or bubbles) have similar movement patterns and, in a self-organized manner, mostly do not cross the borders of their communities. During the pandemic, the patterns became modified because relationships in bubbles became more selective than before the pandemic. Studying the changes of mobility patterns allows us to understand the effectiveness of public health interventions and define the risk in different areas based on the mobility of individuals.

In the following sections, we first provide background information about previous research on functional communities and social segmentation, describe our methodology, and present the results and discussion.

2 Background Information

Mobility patterns have been of interest to economists, sociologists, and geographers for over half a century already. At first, such patterns were known as functional communities, which were introduced in the works by [10, 24, 25, 28] who noticed that as mobility and connectivity of societies increase, boundaries of intensities of economic activities no longer correspond to boundaries of cities and regions. Advancements in transportation, communication and production technologies further decouple functional spaces from physical boundaries and result in discrete patches of economic activities that have their own distinct demographic and economic characteristics.

Historically, functional communities have been defined on the basis of high commuting densities (aka mobility densities). The first attempts to better delineate functional communities were Commuting Zones and Labor Market Areas, first developed in the 1980s [43]. Commuting Zones were defined based on hierarchical cluster analysis and mobility data derived from the Census Bureau's journey to work data. Alternatively, economists have used other delineations, such as Bureau of Economic Analysis Delineations, Federal Communications Commission Delineations, Census Core Based Statistical Areas, and Tong and Plane Delineation, that were based on census data, newspaper circulation movements, and other data [38].

In previous research, communities were defined with the assumption that functional communities are fixed, not dynamic, and do not change often. With the wide availability of open data on transportation and communication networks, understanding of mobility patterns has evolved and allowed researchers to investigate dynamics of mobile communities. Transportation and communication networks nowadays often serve as proxies for capturing the extents of mobile communities. Unlike hierarchically clustered functional communities, mobility patterns are based on human interactions and are derived by means of community detection algorithms such as Louvain. Such communities change in their relative sizes and re-organize during the switches in rhythms in the dynamical systems.

Functional communities have been used in a number of earlier studies which concluded that such communities are valid entities for analysis of various aspects of societies. They can be used for population projection and discussion of economic and social outcomes [23, 27]. [30] investigated the effect of education on local employment growth in functional communi-ties. [16] made use of functional communities in multilevel models to show that both compositional and contextual factors contributed to variations in poverty rates. [2] measured the spatial heterogeneity of teen employ-ment estimates in commuting zones. [21] assessed labor market changes in commuting zones. [34, 36, 37] examined a wide range of demographic characteristics in commuting zones. The role of mobile communities in the context of public health and epidemiology has yet to be understood. Our project attempts to fill this void.

The recent availability of large-scale datasets derived from bank trans-action records, landline, mobile and social media has resurged interest in functional communities and social fragmentation [14, 26, 31]. Apps and so-cial media platforms that trace human movement in geographic space gen-erate mobility data. Each person who moves in space and uses social media for communication, leaves footprints in a form of geospatial coordinates. Geo-located data sources enable direct observation of social interactions and collective behaviors with unprecedented detail. Studies [20] show that online communities do not exist only online: they represent communities in the geographic space.

Mobility maps are beginning to have a broad spectrum of application in policy making and business analytics. They help measure the impact of advertising investments, supply chain optimization, inventory planning, effects of restrictions (e.g., lockdowns), and much more. A number of data startups (e.g., Cuebiq [3], UberMedia [4], SafeGraph [5], Teralytics [6]) are already succeeding at extracting useful nuggets from mobility data and offering powerful solutions to help policy makers and business analysts make better, data-informed decisions about business strategies, public health, logistics, mobile strategies, and community development strategies. Our goal is to investigate how mobility maps can be used for optimizing formation of bubbles.

Unlike in previous studies in economics and sociology, in complex sys-tems, we are looking at functional communities at multiple scales, assuming that each scale captures specific patterns of interactions.

---

[3]https://www.cuebiq.com
[4]https://ubermedia.com
[5]https://www.safegraph.com
[6]https://www.teralytics.net

3 Methods and Materials

Mobility Data

During the COVID-19 pandemic in 2020, mobility data became popular. Mobility datasets capture movements of people from one location to another; such data are the building blocks of functional communities. However, in early stages of the pandemic, mobility data were used only for the analysis of aggregate trends. Aggregate trends of mobility data provided by Google, Facebook, Apple, and Teralytics helped public health officials keep track of the situation with mobility and COVID-19 spread in different parts of the world. Google and Apple provided dynamics plots for changes in mobility in different countries and states [5, 19]. Teralytics, Cuebiq, and Facebook published changes in mobility maps [18, 22, 40], aggregated by counties.

Functional mobile communities were not analyzed. Some companies, however, like SafeGraph, Inc., provided pre-processed anonymized mobility data extracted from smartphone app interactions and invited university researchers to make sense of the mobility patterns. SafeGraph's dataset gave us an opportunity to explore the evolution of functional communities in the US during COVID-19. Coverage of SafeGraphs' datasets is limited to the US and Canada, but data aggregation for Canada is not as complete as in the US.

To anonymize mobility data in its social distancing dataset, SafeGraph aggregated locations of users by census block groups. Census block groups are statistical divisions of counties and states. Block groups contain between 600 and 3,000 people, and are used to present data and control block numbering [42]. Aggregation by block groups prevents data users from reconstructing fine details of footpaths left by individual people. Moreover, it does not give the full path of each individual, rather it gives links only between one main location (or home location) and other locations connected to home, where each location is a census block group. Biases and data cleaning were completed by SafeGraph before the company offered datasets for analysis. So our assumption is that the dataset is free of geospatial and temporal biases.

By its volume, SafeGraph's Social Distancing dataset can be considered Big Data. The dataset includes data for each day from 2019 to 2021; each day's data volume has more than 1GB of data. The dataset has neither geospatial polygons nor centroid coordinates; the dataset has to be enhanced with the Census geospatial datasets. The dataset has nested relationships between block groups, weights (how many people went from home to other locations), and dates.

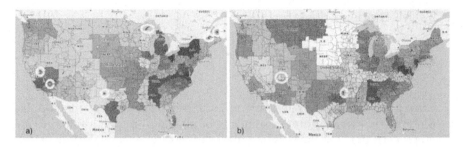

Figure 1: Multiscale changes in communities in two separate weeks: a.) week of March 8-14, b.) week of April 19-25. Colors indicate main communities; black lines distinguish sub-communities; and similar color hues identify communities that belong to the same mega community. Yellow circles show locations of isolated communities.

COVID-19 Cases Data

The Novel Coronavirus (COVID-19) daily data are confirmed cases for affected states reported between 21st January 2020 and 31st December 2020. The data were collected from the reports released by John Hopkins Center for Systems Science and Engineering. The data include confirmed cumulative COVID-19 cases in the US, aggregated by counties.

Data Pre-processing

The original SafeGraph's Social Distancing data comes in the CSV format, grouped by days. The file sizes range from 1 to 2.5 GB. Each file describes individual census blocks and lists links with weights (number of links) to other census blocks that occurred on a specific day. To prepare data for the mobility maps, we first separate all these relationships and describe them as individual objects. Each relationship has a source, a target, the date, and weight of interaction. Daily dataframes are combined into weekly dataframes, grouped, and their relationships are summed. Each census block in each relationship is augmented with central points derived from groups of census blocks that make up a census block group. The datasets are grouped by weeks. This aggregation reveals better patterns in data.

Mobility Network

The relational data from SafeGraph's Social Distancing dataset is used for the mobility network. In the mobility network, nodes represent census block groups. Edges denote the movement of an individual from one location (node) to another one. Here, edges' weights correspond to the number of people who travel between the two census block groups.

Community Detection Algorithm

Mining communities from data is commonly associated with the discovery of tightly connected user subgroups linked by a specific context. We analyze the presence of communities by applying the Louvain method [12] with modularity optimization [15] to the mobility network. Communities refer to the regions in which nodes are more connected to each other than the rest of the network. In the Louvain method, in an iterative process, nodes move to neighboring communities and join them to maximize modularity (M). Modularity is a scalar value $1 < M < 1$ that quantifies how distant the number of edges inside a community is from those of a random distribution. Values closer to 1 represent better detected communities.

We run community detection algorithms 3 times for each time frame. During the first run, we detect large communities, then we select relationships within each large community and detect sub-communities. Finally, to detect mega communities, we select relationships among sub-communities, filtering out relationships inside the communities. Mega communities are used for defining color hues on the maps. Typically, the number of mega communities is not large, ranging from 5 to 6. For each group, we are trying to assign similar color hues over time, so that sub-communities in Florida and Alabama have blue shades, in the Mid-West grey, in the North-East purple, in the South pink, and in the West from yellow to brown.

Doing community detection analysis at multiple resolutions is crucial for understanding behavior at multiple scales. While at the country level, dynamics of positive COVID-19 tests might appear to be increasing, at the sub-communities level, spatial heterogeneity in dynamics can be observed: some communities might have faster growing cases than others. The higher the resolution, the more accurately we can assess the local risk of infection.

Dot and Polygon Maps

Census block groups, geospatial units of analysis used by SafeGraph, can be represented as polygons or as central points. We used both types of maps in our analysis. Dot maps are quick and easy to explore; polygon maps are more convenient for analysis. To produce polygon maps, we assigned codes of large communities to each polygon and then dissolved polygons based on their names and assigned colors. Then we overlaid sub-communities over main communities, highlighting only their borders.

4   Results and Discussion

To date, we have 16 mobility maps that show how communities changed over time from January to the end of May and some weeks in the fall of 2020. The earlier maps are in the dot format; later maps display polygons.

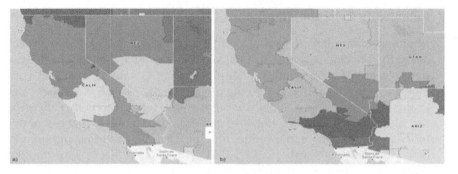

Figure 2: Communities do not fit into the state boundaries at different time periods: a.) February 23-29; b.) April 5-11. Colors indicate main communities; yellow boundaries show state borders.

Maps show the evolution of communities over time. In this section, we summarize some of the insights we gained from these maps.

Maps over time demonstrate that we all lived in social bubbles before and during the pandemic. Bubbles went through numerous transformations during the pandemic due to lockdown, shelter-in-place and other mitigation policies. Some bubbles, however, were more affected by these policies than others. Compare the following two maps in Figure 1.

Colors on these maps indicate main communities (bubbles); black lines separate sub-communities; and similar color hues identify communities that belong to the same mega community. The map on the left (Figure 1.a) shows communities from March 8 to 14, and to the right (Figure 1b.) from April 19 to 25. The first map shows mobility patterns just before the lockdown, and the second at the end of the 5th week of the national lockdown. Mega community membership is very different in these time periods. Before the lockdown, Michigan and Ohio were grouped together with New York, while during the lockdown they merged with the group of mid-western states. This change suggests that there was greater than usual movement between New York City/New England, and Michigan/Ohio. Before people settled in for shelter-in-place, they moved in space. Another difference between these two maps is that, on the first map (Figure 1a.), Florida is in a community with New York, which suggests that there was a significant movement between these communities too. Indeed, it was the time of spring break and vacations; according to news reports at that time, many students traveled to Florida. Another three communities that changed memberships were Wyoming, Colorado, and New Mexico, perhaps also due to increased traffic flows. Small communities in these maps are important too; they explain spatial heterogeneity in movements, even if none is present at the higher levels of detail.

In addition, it is important to note that the map in Figure 1a. appears to have more isolated communities (these communities are marked with

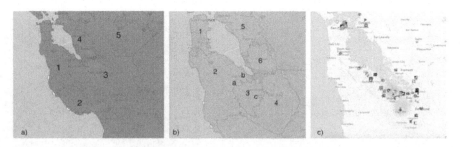

Figure 3: Temporal changes in sub-communities in Silicon Valley: a.) sub-communities in the pre-lockdown week (February 23-29); b.) sub-communities during the lockdown week (April 5-11); c.) heatmap of high-tech companies [7] that partially explains groupings in the previous two maps. Colors indicate main communities; black boundaries show borders of subcommunities. Letters and numbers in a. and b. refer to distinct subcommunities or locations.

yellow circles). Isolated communities are separated from the original communities by distance. Some of these isolates are related to trips to national parks, others to locations of marine corps training sites, air bases, or Indian reservations.

Both maps have large numbers of main communities (62 and 64 respectively), compared with 48-54 during the pre-COVID weeks. This suggests that the lockdown measures helped to reduce unnecessary movements and the spread of the disease.

Mobile communities do not conform with administrative regions. Consider zooming in on Western states (Figure 2): California, Nevada and Arizona. Maps in Figure 2 show that these communities do not fit into the state boundaries (state borders are shown as yellow lines) neither before the lockdown in February 23-29 (Figure 2a.), nor during the lockdown in April 5-11 (Figure 2b.), even though the number of communities during the lockdown has significantly increased. California has 6 or 7 communities, Nevada 3 and Arizona 4, respectively. The majority of these communities go across the state borders. While some communities vary in size a little bit, others often merge with other communities (e.g. Northern California and Nevada in Figure 2a. or a large community in Southern California and Arizona). Some of these mergers can be explained by the location of nature zones or national parks at the fringes of these communities. With greater visits to parks, the divide between communities dissolves.

The maps allow us to observe the breakup of the communities and sub-communities into smaller ones over the course of the pandemic. For example, in Figure 3 we show how the Silicon Valley community looks before the lockdown (Figure 3.a), during the lockdown (Figure 3.b) and how the locations of high tech companies can explain these changes (Figure 3.c). Before the lockdown, we can clearly see 5 large sub-communities. After the lockdown was imposed, during April 5-11, the number of sub-communities has

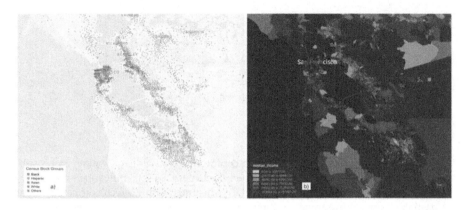

Figure 4: Socio-demographic maps of Silicon Valley area: a.) ethnic map; b.) income map. Colors in 4a. indicate prevalent ethnicities at each census block group. Colors in 4b. indicate income ranges at each census block group.

increased (we highlighted 6). Before the lockdown, the San Francisco area was grouped together with San Mateo and Redwood City (see 1 in Figure 3a.); there were no natural boundaries between them. Palo Alto, Mountain View, and Stanford University formed a large sub-community with San Jose (see 3 in Figure 3.a), also without any boundaries. During the lockdown, San Francisco separated from San Mateo and Redwood City. Stanford University detached from San Jose and merged with Redwood City (see b. in Figure 3b.). The Stanford University hospital and Kaiser Permanente Santa Clara Medical Center formed their own detached sub-communities (see a. and c. in Figure 3b.). The pre-lockdown Oakland/Fremont community (see 4 in Figure 3a.) split into two communities in April (see 5 and 6 in Figure 3b.).

   All of the above changes in bubbles are associated with the mobility patterns, ethnic composition and income inequalities. Maps in Figure 4 show ethnic and income associations in the Silicon Valley. Compare these maps with maps in Figure 3. It is evident from Figure 4 that San Jose (earlier marked as 4 in Figure 3b.) is strikingly different ethnically and income-wise from the community 3 in Figure 3b. which includes Mountain View, Sunnyvale, and Palo Alto. Similarly, other mobile sub-communities have not only natural breaks in mobility patterns, but also ethnic and income contrasts.

5   Grey areas

Maps have a few urban areas that do not share any mobility data. A pair of census block groups in Hidden Hills City, CA, (see Figure 5) is one such grey area. Hidden Hills is a city of celebrities. It is a city where Kim Kardashian,

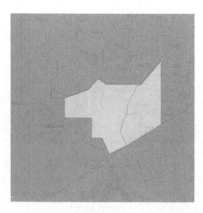

Figure 5: Grey census block groups. Orange color around the grey polygon indicates main community.

Kanye West, Drake, Justin Bieber, and Khloe Kardashian live or used to live. SafeGraph does not have any mobility data from these block groups.

Finally, in the next set of images (Figure 6), we show cumulative COVID-19 cases over the two week period prior to April 11, 2020. Cases are aggregated by main communities (Figure 6.a) and sub-communities (Figure 6.b). Raspberry color identifies bubbles with high contagion rates, green low, and light-pink bubbles in between. The legends of the maps are slightly different because main communities are much larger than sub-communities. While large communities are considered green when they have fewer than 20 active cases, subcommunities may have only 10 cases to be considered green. The map in Figure 6.a shows that the majority of large communities are highly infected and are not safe. Infection rates in their corresponding sub-communities in Figure 6.b, however, are not spatially heterogeneous and may include low risk areas too. The Washington state bubble, for example, is not all high risk as shown in Figure 6.a: it includes green, red, and white bubbles in Figure 6.b. For travel inquiries, the map of subcommunities gives information about safe bubbles where people can go for hikes or short-distance travel amid the pandemic. For business recovery, the levels of contagion in sub-communities should be directly translated into mitigation policies. Changes in shapes of communities and sub-communities overtime give information about increases or decreases of movements to and from communities. While some bubbles are stable over time, others go through different transformations at the edges, or break down into smaller sub-communities, suggesting that dynamics in those communities or sub-communities is changing and may lead to higher infection rates within a short period of time.

Maps in Figure 6 also inform policy makers about optimal alliances for travel and business. Mega communities give recommendations on how to form bubbles between states (see yellow borders on both maps). Mega

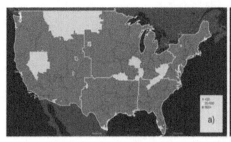

Figure 6: COVID-19 cases over two weeks prior to April 11, 2020 aggregated by: a.) main communities; b.) sub-communities. Colors in 6a. indicate contagion levels in main communities; yellow lines show boundaries of megacommunities. Colors in 6b. indicate contagion levels in sub-communities; thick black lines show borders of main communities, yellow boundaries of megacommunities.

communities include sub-communities with different levels of contagion, letting people have places for travel and safe business within the same large bubble.

For contact tracing, understanding these multiscale social bubbles is also very important. It gives information about possible contagion risks. For example, contrary to common knowledge that New York state was highly contagious in April of 2020, coming from New York state was not always high risk. Upstate New York, for example, was not in the community with New York City, where contagion level was extremely high. One could have even come from a green area in upstate New York, depending on which bubble he or she was coming from.

6   Conclusion And Future Work

In this chapter, we demonstrated how social bubbles evolved in the United States during the first national lockdown due to the COVID-19 pandemic in the spring of 2020. The bubbles are characterized by spatial heterogeneities in terms of their shapes and contagion levels: they vary by scales and do not correspond to administrative boundaries. Besides mobility patterns, we also identified high and low risk areas for contagion. These patterns provide information to policy makers on how to impose travel restrictions and quarantine policies that are minimally disruptive for social and economic activities, and on how to analyze the effectiveness of contact tracing.

Along with data and analytical analysis, we are also considering predictive simulation models for epidemics to understand the evolution of COVID-19 and to plan efficient management strategies. Since the 1920s, compartment models in the form of differential equations have been proposed to study infectious disease dynamics and human-to-human transmission of diseases [29]. These models help to understand the effectiveness of interventions considered in the body of the models. These models divide a population of N individuals into different stages such as the susceptible, infected and recovery stages in the SIR model [9]. In the simplest form, models consider no structure for the societies and populations are assumed completely mixed. However, recent studies [1,7,8,44] show that geographical heterogeneity in the populations and individuals' movement patterns have serious impacts on the dynamic of diseases. In spatial metapopulation models, simulations of disease spread run on a network of sub-populations that are connected through the movements of individuals between areas. We are taking our maps in this direction next.

Bibliography

[1] Adiga, A., Dubhashi, D., Lewis, B., Marathe, M., Venkatramanan, S., Vullikanti, A.: Mathematical models for COVID-19 pandemic: a comparative analysis. Journal of the Indian Institute of Science, pages 1–15. (2020)

[2] Allegretto, S., Dube, A., Reich, M.: Spatial Heterogeneity and Minimum Wages: Employment estimates for teens using cross-state commuting zones. Institute for Research on Labor and Employment: Working Paper Series. (2009). http://escholarship.org/uc/item/1x99m65f

[3] Allen, J., Szekely, P.: New York, Connecticut and New Jersey to quarantine visitors from states with high COVID-19 infection rates. Reuters, New York (2020)

[4] Amini, A., et al.: The impact of social segregation on human mobility in developing and industrialized regions. EPJ Data Science, 3 (1), 1–20. doi:10.1140/epjds31 (2014)

[5] Apple, Inc. Mobility Trends Reports (2020) url-https://covid19.apple.com/mobility

[6] Appleton, N. S.: The Bubble: A New Medical and Public Health Vocabulary for COVID-19 Times (2020) urlhttp://somatosphere.net/2020/the-bubble.html/

[7] Arandiga, F., Baeza, A., Cordero-Carrion, I., Donat, R., Mart, M. C., Mulet, P., Yanez, D. F.: A spatial-temporal model for the evolution of the covid-19 pandemic in spain including mobility. Mathematics, 8(10):1677 (2020)

[8] Balcan, D., Colizza, V., Goncalves, B., Hu, H., Ramasco, J. J., Vespignani, A.: Multiscale mobility networks and the spatial spreading of infectious diseases. Proceedings of the National Academy of Sciences, 106(51):21484-21489 (2009)

[9] Barlow, N. S., Weinstein, S. J. (2020) Accurate closed-form solution of the SIR epidemic model. Physica D: Nonlinear Phenomena, 408:132540.

[10] Beale, C. L., Fuguitt, G. V.: The new pattern of nonmetropolitan population change. In Social Demography(pp. 157-177). Academic Press (1978)

[11] Berger, E., Reupert, A.: The COVID-19 pandemic in Australia: Lessons learnt. Psychological Trauma: Theory, Research, Practice, and Policy, 12(5), 494 (2020)

[12] Blondel, V. D., Guillaume, J., Lambiotte, R., Lefebvre, E. Fast unfolding of communities in large networks. Journal of Statistical Mechanics: Theory and Experiment. 2008 (10): P10008. urlarXiv:0803.0476. Bibcode:2008JSMTE..10..008B urldoi:10.1088/1742-5468/2008/10/P10008 S2CID 334423 (2008)

[13] Buchel, O., Ninkov, A., Cathel, D., Bar-Yam, Y., Hedayatifar, L.: Strategizing COVID-19 Lockdowns Using Mobility Patterns. arXiv preprint url-https://arxiv.org/abs/2012.03284 (2020)

[14] Calabrese, F., Dahlem, D., Gerber, A., Paul, D., Chen, X., Rowland, J., et al.: The connected states of America: Quantifying social radii of influence. In: 2011 IEEE Third International Conference on Privacy, Security, Risk and Trust and 2011 IEEE Third International Conference on Social Computing (pp. 223-230). IEEE (2011, October)

[15] Clauset, A.; Newman, M. E. J., Moore, C.: Finding community structure in very large networks, Physical Review E2004, 70, 066111 (2004)

[16] Cotter, D. A.: Poor people in poor places: Local opportunity structures and household poverty. Rural Sociology, 67(4), 534–555 (2002)

[17] Craig, T., Dennis, B.: Governors form groups to explore lifting virus restrictions; Trump says he alone will decide. The Washington Post (April 13, 2020)

[18] Cuebiq, Inc.: Shelter-In-Place Analysis. urlhttps://www.cuebiq.com/visitation-insights-sip-analysis/ (2020)

[19] Google, Inc.: COVID-19 Community Mobility Reports. url-https://www.google.com/covid19/mobility/ (2021)

[20] Hedayatifar, L., Rigg, R. A., Bar-Yam, Y., Morales, A. J.: US social fragmentation at multiple scales. Journal of the Royal Society Interface, 16(159), 20190509 (2019)

[21] Dupor, B., McCrory, P.: A cup runneth over: Fiscal policy spillovers from the 2009 recover act. Federal Reserve Bank of St. Louis: Working Paper Series. urlhttps://onlinelibrary.wiley.com/doi/epdf/10.1111/ecoj.12475 (2014)

[22] Facebook, Inc.: Movement Range Maps urlhttps://data.humdata.org/dataset/movement-range-maps (2020)

[23] Fowler, C. S., Rhubart, D. C., Jensen, L.: Reassessing and revising commuting zones for 2010: History, assessment, and updates for US 'labor-sheds' 1990–2010. Population Research and Policy Review, 35(2), 263-286 (2016)

[24] Frey, W. H., Speare, A.: Metropolitan areas as functional communities. In Metropolitan And Nonmetropoolitan Areas: New Approaches To Geographic Definition. Working Paper No. 12. Washington DC: US Census Bureau. (1995)

[25] Garreau, J. Edge city: Life on the new frontier. Anchor. (1992)

[26] Gonzalez, M. C., Hidalgo, C. A., Barabasi, A. L.: Understanding individual human mobility patterns. nature, 453(7196), 779-782 (2008)

[27] Hildner, K. F., Nichols, A., Martin, S.: Methodology and assumptions for the mapping America' s futures project. Washington D.C. (2015)

[28] Kasarda, J. D.: Jobs, migration, and emerging urban mismatches. Urban change and poverty, 148-198 (1988)

[29] Kermack, W. O., McKendrick, A. G.: A contribution to the mathematical theory of epidemics. Proc. R. Soc. Lond., A115700, 721 (1927)

[30] Killian, M. S., Parker, T. S.: Education and local employment growth in a changing economy. Education and Rural Economic Development: Strategies for the 1990s, 93-121 (1992)

[31] Lazer, D., Pentland, A., Adamic, A., ARAL, S., Barabasi, A.-L., Brewer, D., Christakis, N. et al.: Computational Social Science. Science 323 (5915): 721–3 (2009)

[32] Long, N. J., Aikman, P. J., Appleton, N. S., Davies, S. G., Deckert, A., Holroyd, E., et al.: Living in bubbles during the coronavirus pandemic: insights from New Zealand: A Rapid Research Report (May 13, 2020)

[33] Noar, S. M., Austin, L.: (Mis) communicating about COVID-19: Insights from Health and Crisis Communication. Health communication, 35(14), 1735-1739 (2020)

[34] Martin, S., Astone, N. M., Peters, H. E., Pendall, R., Nichols, A., Hildner, K. F., Stolte, A.: Evolving patterns in diversity. Washington D.C.: Urban Institute (2015)

[35] Menezes, T., Roth, C.: Natural scales in geographical patterns. Scientific reports, 7(1), 1-9 (2017)

[36] Nichols, A., Martin, S., Astone, N., Peters, H.: The labor force in an aging and growing America. Washington, DC: The Urban Institute (2015)

[37] Pendall, R., Martin, S., Astone, N., Nichols, A.: Scenarios for regional growth from 2010 to 2030. Washington, DC: The Urban Institute (2015)

[38] Penn State: Penn State Commuting Zones / Labor Markets data repository. https://sites.psu.edu/psucz/ (n.d.)

[39] Shalizi, C. R., Camperi, M. F., Klinkner, K. L.: Discovering functional communities in dynamical networks. In ICML Workshop on Statistical Network Analysis (pp. 140-157). Springer, Berlin, Heidelberg (2006, June)

[40] Teralytics, Inc.: Mobility and COVID-19 urlhttps://www.teralytics.net/mobility-and-covid-19/ (2020)

[41] Trnka, S., Davies, S. G.: COVID-19, New Zealand's bubble metaphor, and the limits of households as sites of responsibility and care. COVID-19: Volume I: Global Pandemic, Societal Responses, Ideological Solutions (2020)

[42] United States Census Bureau: Glossary urlhttps://www.census.gov/programs-surveys/geography/about/glossary.html (2019)

[43] United States Department of Agriculture (USDA): Economic Research Service: Commuting Zones and Labor Market Areas. urlhttps://www.ers.usda.gov/data-products/commuting-zones-and-labor-market-areas/ (2019)

[44] Venkatramanan, S., Chen, J., Fadikar, A., Gupta, S., Higdon, D., Lewis, B., Marathe, M., Mortveit, H., Vullikanti, A.: Optimizing spatial allocation of seasonal in uenza vaccine under temporal constraints. PLoS computational biology, 15(9):e1007111 (2019)

[45] WEGO Travel: What is a Travel Bubble? Here's Everything You Need to Know About the Buzzy New Term in Travel. urlhttps://blog.wego.com/whats-a-travel-bubble/ (January 18 2021)

[46] Xiang, F., Tu, L., Huang, B., Yin, X.: Region partition using user mobility patterns based on topic model. In: 2013 IEEE 16th International Conference on Computational Science and Engineering, pp. 484-489. IEEE (2013, December)

COVID-19 - The Swedish Experiment: A Data Driven Analysis of the Swedish Corona Strategy

Jan Brugård*

Wolfram MathCore AB,
Teknikringen 1E, 583 30 Linköping, Sweden
http://www.wolframmathcore.com/
*Corresponding Author: janb@wolfram.com

When the corona virus started to spread around the World most countries opted for strict rules and regulations. Including things like lockdown, mandatory masks, and fines. Sweden, on the other hand, opted for recommendations rather than rules. Partly because of tradition, partly due to limitations given by Swedish law. As this approach was different from most it was soon named *the Swedish experiment*. Living in Sweden, I thought it would be interesting to see how the strategy was performing. This article is an attempt to do a data driven analysis of the Swedish measures to get the corona pandemic under control.

Method

The Wolfram Language [1] and public available data has been used to study the impact of measures taken by the Swedish authorities in order to control the corona pandemic, using basic data visualization.

The study was done in four different parts, in May 15, June 15, September 14, 2020, and January 28, respectively. As time progressed, I learned more about pandemics in general and the corona pandemic specifically. That said, except for minor bug fixes, I have opted not to update any analysis done or conclusions made during the different stages, to give a better sense for how my knowledge, hopefully, evolved over time. This also means that conclusions made at one point in time might not be completely in line with conclusions made at others. The original analysis and corresponding code for the three initial parts can be found in the Wolfram Community post *COVID-19 - the Swedish experiment - is it working* [2].

1 Analysis from May 15, 2020

The first known case of COVID-19 in Sweden, a traveller returning from China, was reported on January 31 [3]. Five days later, on February 5, the

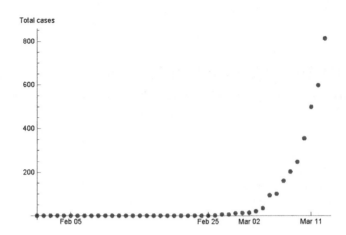

Figure 1: Accumulated number of known cases of COVID-19 in Sweden, during January 31 to March 9, 2020.

Swedish authorities announced that people returning from China should contact healthcare [4] in case they showed any symptoms.

It took until February 25 before authorities upgraded the risk in Sweden to *high* [5]. One day later, on February 26, the second case was reported [6]. This time it was a person returning to Gothenburg from Italy. The day after, five new cases were reported [7]. One of them had been infected in Italy, one in Germany, and one in Iran. Finally, two of them had been found through tracing and had been infected by the one of the other known cases. The person coming from Iran was the first known case in Stockholm. Fig. 2 shows how the initial period looked like.

On March 2, the Public Health Authority of Sweden requested that Iran Air's flights from Iran should be cancelled [8] to prevent spread from Iran. The same day, they announced that the risk of encountering new cases in Sweden was increased to *very high* [9]. The following day, they recommend testing [10] all returning persons from northern Italy if they showed any symptoms within 14 days. On March 6, the Foreign Ministry of Sweden issued a recommendation to avoid travels to northern Italy [11]. At the same time the number of reported cases had started to increase.

At this time, a total of 800 had been tested positive. The first death in COVID-19 in Sweden was reported on March 11. After this the number of deaths started to slowly grow as seen in Fig. 2.

On March 9, a total of 250 people had been tested positive, and the day after the Public Health Authority stated that they could now see a societal spread in Sweden [12]. They made it clear that anyone, that showed any symptoms should limit their social interactions. Following, on March 11, a recommendation to limit public gatherings [13] to a maximum of 500 persons was issued. The same day the WHO declared Corona as a

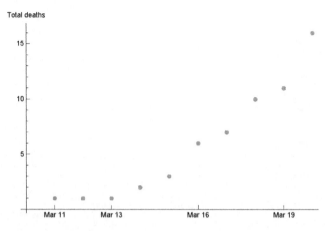

Figure 2: Accumulated number of deaths with COVID-19 in Sweden, during the beginning of the pandemic.

pandemic [14] and on March 13, the Public Health Authority declared a Sweden had entered a new stage in the fight against the pandemic [15]. At this stage, the focus was on delaying the spread and protecting the elderly. Sweden's chief epidemiologist, Anders Tegnell, stated: *"The important thing is that everyone takes their responsibility and stay at home when they are not healthy and for safety two day after one getting healthy and not shown any symptoms at all"*.

One and a half month after the first case in Sweden, on March 16, persons over 70 years were recommended to restrict their social contacts as much as possible and a recommendation to work from home if possible was issued [16]. On March 17, High Schools and Colleges were recommended to switch to remote teaching [17]. The following days new recommendations were added, including travel recommendations [18] (March 19), limitations for restaurants [19] (March 24), and further restrictions on public gatherings, limiting them to a maximum of 50 people [20] (March 27). As you probably noted by now, the measures taken, except the limitations on gatherings, were mainly recommendations.

At this time, media around the World had started to write about *the Swedish experiment*, as in these articles: *Sweden bucks global trend with experimental virus strategy* (Financial Times, March 25) [21], *In the Coronavirus Fight in Scandinavia, Sweden Stands Apart* (New York Times, March 28) [22], *Sweden goes against the current: full means of transport and open offices* (Repubblica, March 26) [23], and *the Swedish exception* (El País, April 4) [24].

During April, there were only minor adjustments to the recommendations [25]. Fig. 3 shows how *the Swedish experiment* had worked that far.

The periodic trend seen is explained by under reporting during weekends and holidays. There were more than 3,000 fatalities reported at this

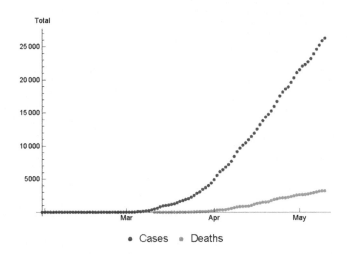

Figure 3: Accumulated number of known cases of and deaths with COVID-19 in Sweden, during the first months of the pandemic.

point. This was among the highest reported number of deaths per capita in the World. Compared to the known cases the rate was also very high. With a case fatality rate of more than 10%, it looked like Sweden was not doing that well. But was that the whole truth?

The NY Times has a continuously updated article in which they compare the excess deaths and official COVID-19 numbers for several countries [26], including Sweden (new countries are added over time). If there is a big difference between these, then it is likely some problem in the COVID-19 reporting. For instance, Ecuador (where I grew up) had reported 1,561 COVID-19 fatalities during March to April. However, they had 10,100 fatalities more than normal during the same period. In contrast Sweden reported 2,996 COVID-19 deaths during March 16 to May 3, compared to 3,300 excess deaths. In other words, the reliability of reported numbers seemed to vary a lot from one country to another. Fig. 4 shows the full table from NY Times as of May 2020.

So, let us look a bit more at the data from Sweden again. On May 11, Statistics Sweden reported that the excess deaths in Sweden were going down [27] (but with variations though out the country). This data is provided and continuously updated by Statistics Sweden as an Excel spreadsheet [28], which I downloaded on May 14. In Fig. 5 the death rate is compared with previous years.

The sudden drop for some years at the end of February is because it is the leap day, i.e., February 29. The drop near day 130 for 2020 was likely due to a lag in reporting. For the first 80 days, i.e., until March 20, the death toll was around average, or actually slightly less than average. Thus, when the media started to write about *the Swedish experiment* death tolls were still

Where we found higher deaths than normal

AREA	PCT ABOVE NORMAL	EXCESS DEATHS	REPORTED COVID-19 DEATHS	DIFFERENCE
U.K. Mar. 14 - May 1	67%	53,300	36,586	16,700
Italy March	49%	24,600	13,710	10,900
Ecuador March - April	84%	10,100	1,561	8,500
Spain Mar. 16 - May 3	60%	31,500	25,213	6,300
France Mar. 16 - Apr. 26	44%	28,500	22,708	5,800
New York City Mar. 11 - May 9	277%	24,200	19,931	4,300
Netherlands Mar. 16 - Apr. 26	50%	8,700	4,463	4,200
Five cities in Brazil Mar. 30 - Apr. 26	39%	5,300	2,122	3,100
Jakarta March - April	62%	3,300	381	2,900
Peru March - April	21%	3,600	1,051	2,600
Istanbul Mar. 9 - Apr. 26	36%	3,100	1,683	1,400
Germany Mar. 16 - Apr. 12	5%	4,000	2,661	1,300
Moscow April	17%	1,700	642	1,000
Portugal Mar. 16 - Apr. 12	15%	1,300	504	800
Switzerland Mar. 16 - May 3	24%	2,000	1,483	600
Austria Mar. 16 - Apr. 26	11%	1,000	541	500
Sweden Mar. 16 - May 3	27%	3,300	2,996	300
Denmark Mar. 16 - May 3	5%	300	492	<0
Israel March	Normal	-200	20	<0
Norway Mar. 16 - Apr. 26	Below normal	<100	198	<0
South Africa Mar. 18 - Apr. 28	Normal	-2,900	93	<0
Belgium Mar. 16 - Apr. 19	31%	3,300	6,336	<0

Note: Excess deaths are estimates that include deaths from Covid-19 and other causes. Reported Covid-19 deaths reflect official coronavirus deaths during the period when all-cause mortality data is available, including figures that were later revised. Reported Covid-19 deaths in Istanbul are estimated based on the government's statement that 60 percent of the country's cases are in the city.

Figure 4: Table from NY Times comparing the number of recorded deaths in COVID-19 with the excess deaths in several different countries.

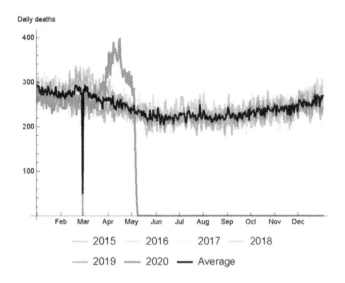

Figure 5: Daily number deaths in Sweden from 2015 to April 2020.

quite normal. However, after that it is obvious that Sweden had a death rate that was clearly above average. As seen, at the end of this period, death tolls were getting closer to normal again. Note that there is always a couple of days delay in the reporting, so the last couple of points are less reliable.

An interesting thing to look at is how we compare the total for each year. Fig. 6 shows the total per 100,000 population for the period January 1 to May 4 each year. Note that the leap years, i.e. 2016 and 2020, have one extra day compared to the others.

2020 is slightly lower than the worst year (2015), but as seen the difference is not striking. However, if you go back to the graph for daily number of deaths in Sweden, Fig. 5, it is clear that Sweden has had a higher death rate than normal in April this year.

Verdict

So, if the total number of deaths is fairly similar to previous years, does that mean that *the Swedish Experiment* is working? Not necessarily, as it all depends on how far the pandemic has come. Do we have a lot of undiscovered cases and were we getting close to herd immunity? At this point, it is probably still a long way to go, and until we have reliable numbers on the total number of infected, it is not possible to make a verdict.

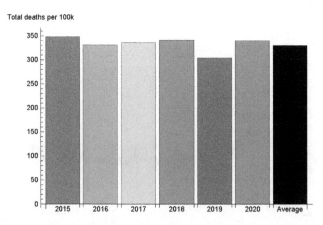

Figure 6: Total number of deaths per year and 100,000 population in Sweden from 2015 to 2020.

2 Analysis from June 15, 2020

A month after I published the original post above, I made a second analysis, which is presented in this section.

On May 12, the first results from ongoing antibody tests were presented by the Public Health Authority [29]. The analysis was based on tests from the last week of April, and 3.7% - 7.3% of the tests showed antibodies (as it takes some time to get antibodies this shows the minimum level of people that had been infected to date). Tests also showed that while 6.7% in the age group 20 - 64 years had antibodies, only 2.7% of the elderly and 4.7% of the young had it. Indicating that protection of elderly had worked to some extent and that, as expected, letting kids attend school had not resulted in a lot of kids becoming infected. Later, on May 29, it was decided to allow sports competitions from June 14 [30], but without spectators. The potential effect of this was yet to be seen of course. Fig. 7 shows an updated version of the daily number of deaths in Sweden, using data from June 15.

The trend of decreasing number of deaths had continued, but Sweden was still on high levels. Normalizing the annual number of deaths with respect to the population should make things more comparable. Fig. 8 shows the number of deaths per 100,000 population.

As could be expected, by this time 2020 had surpassed 2015 as the worst of the last 6 years. Fig. 9 shows that 2015 had a higher death rate during the first quarter (due to a hard influenza in 2015 and a mild in 2020), however, 2020 caught up during the second quarter. The impact of the pandemic was comparable to the difference between a hard and a mild influenza in other words. At least to this stage and without considering other factors such as economy, health, and employment.

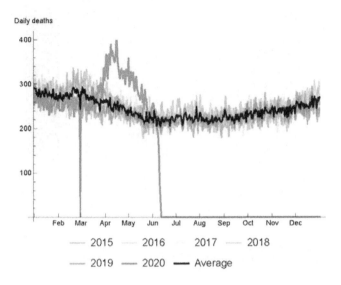

Figure 7: Total daily number of deaths in Sweden 2015 to May 8 2020.

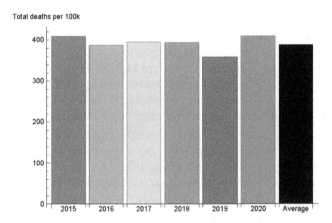

Figure 8: Total number of deaths per 100,000 population in Sweden during the first 150 days of each year.

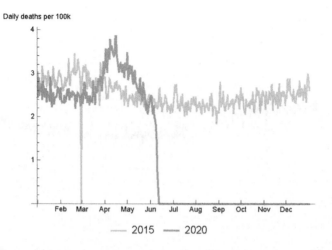

Figure 9: Comparing total number of daily deaths in Sweden in 2020 (up to June 10) with 2015.

Verdict

Remember that a delay in reporting affects especially the last week, but to some extent previous weeks too. Still, at this point, it looked like the infection was heading in the right direction. However there should be a couple of interesting weeks to see if Sweden was back to something close to normal or if it started to increase again.

To this point, Sweden had been hit harder than most other countries. Whether this was due to a different strategy or other factors is hard to say. On the other hand, the number of excess deaths was comparable to the difference between a hard and a mild influenza season. Put in that perspective, and not accounting for the impact on e.g. education, economy, and jobs, my conclusion was that the effects were reasonable. However, yet again, we were probably early in the pandemic, so it all depended on what happened next.

3   Analysis from September 14, 2020

When I wrote the previous section in June 15, I was planning updates every month. But not too much happened in Sweden for a while, so having monthly updates felt as a bit too much.

Most Swedes have vacation around July and August, so on September 14 most had been back to work or school for a while. This is typically a time of the year when many Swedes catch a cold as they start to meet in new groups after the vacations, therefore it is interesting to see if there were any

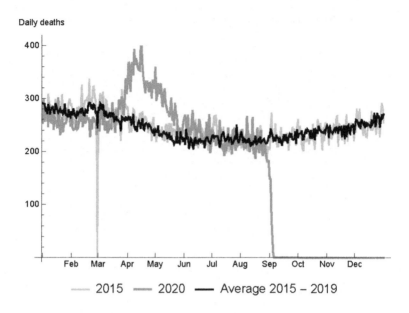

Figure 10: Comparing total number of daily deaths in Sweden in 2020 (up to September 10) with 2015 and the last five years average.

signs of COVID-19 increasing again. Fig. 10 shows an updated version of the total number of deaths in Sweden to this point.

In Fig. 8, I plotted deaths per 100,000 population. However, at this point, I decided not to normalize when comparing back in time, as I learnt that most of the 5% increase in population from 2015 to 2020 was in younger age groups (including many refugees). Thus, it could be expected that the number of deaths should be kept fairly constant. Nevertheless, the graph in Fig. 10 shows that Sweden was back on pretty much normal levels in September 14.

As shown in Fig. 11, the death rate was continuously decreasing over the summer. The attentive reader might notice a small dip in accumulated deaths at the end of August, which is confusing of course. However, some Swedish regions had reported cases in a slightly different way than the official, which was then corrected at the end of August.

Protecting elderly

One of the main criticisms regarding Sweden's handling of COVID-19 has been a failure of protecting the elderly. Therefore, on June 12, the Public Health Authority published examples of successful geriatric care as an inspiration to other care centers [32]. Furthermore, on June 17, new guidelines for PCR testing and infection tracing in elderly care were published [33].

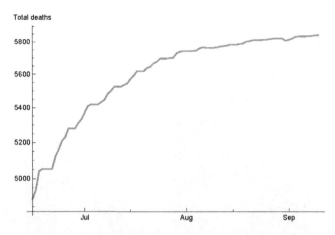

Figure 11: Accumulated number of deaths with COVID-19 in Sweden from June 15 to September 10 2020.

In order to get an indication whether these changes were having any effect, I downloaded the latest data again on September 7 [28]. Fig. 12 shows the change in number of deaths per age group and gender for the first 243 days of 2019 and 2020, respectively. At this point, there was a clear increase for all age groups and genders, even though the increase for females, 64 years or less, was small.

Now, to the question, did the changes in June 12 have any effect? As it typically takes a few weeks from infection to possible death I chose to compare 60 days from July 3 - August 31, with the prior 60 days, see Fig. 13. For males there seems to be no difference in the decrease between the different age groups. On the other hand, for females, there is a clear difference between age groups. There are of course many other factors that might influence, but I would say that, for males, it is not obvious that elderly was protected better than before. At least not on this short term. However, for females the comparison indicates that the changes might have had a positive effect for the elderly.

Comparing regions

While most other countries were changing recommendations and regulations related to COVID-19 on a regular basis, Sweden continued with few changes. A new law for restaurants and cafés [34] was introduced on July 1, giving them increased responsibility for taking infection control measures (which were outlined the day after by the Public Health Authority [35]), on July 30 they repeated that people should continue to work from home if possible [36], new recommendations for choirs [37] were given on August 13, finally on August 31, children and youngsters with symptoms were

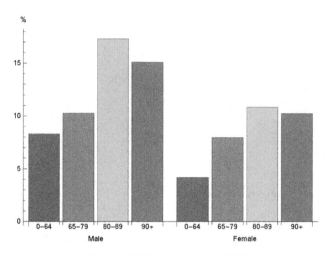

Figure 12: Increase in death rate in 2020 compared to 2019 (using the first 243 days).

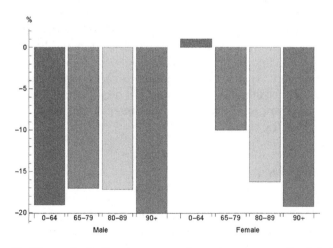

Figure 13: Change in death rates per gender and age group after July 3 2020.

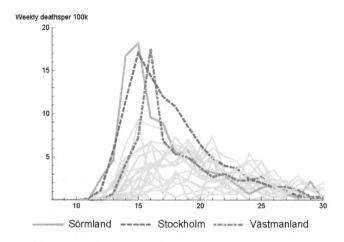

Figure 14: Weekly number deaths with COVID-19 per 100,000 persons in different Swedish regions, highlighting Sörmland, Stockholm, and Västmanland.

recommended PCR testing [38] in order to get back to school faster. Other than that, there were no substantial changes. Thus, the strategy continued to be the same, but some minor implementation details were adjusted.

Except some minor differences (for instance in short term travel recommendations) all Swedish regions had followed the same recommendations and had been handled the same way, therefore it is interesting to compare outcome between them, see Fig. 14.

Together with neighbouring Sörmland and Västmanland, Stockholm had a fairly similar curve during this part of the pandemic, and they were clearly ahead of the other regions. It is easy to see that the development had varied a lot between regions, i.e., despite having the same strategy variations was big. The principal explaining factor seemed to be that regions were the pandemic arrived earlier, and therefore grew much before restrictions were put in place got a fast growth that took quite a bit of time to get under control.

In the beginning rather few tests were made, which can clearly be seen if we look at number of cases per county, see Fig. 15 and compare this with the number of deaths shown in Fig. 14. This illustrates how hard it is to say anything based on only number of known cases.

Verdict

As the situation in Sweden as well as in other countries around the world had changed, international media's view on *the Swedish experiment* had also shifted somewhat. Most media articles, like the (very interesting) article *Anders Tegnell and the Swedish Covid experiment* [39] on September 11 in

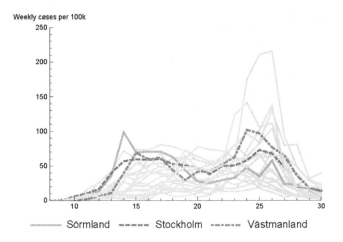

Figure 15: Weekly number of known cases of COVID-19 per 100,000 population in different Swedish regions, highlighting Sörmland, Stockholm, Västmanland, and Västra Götaland.

Financial Times were neutral. Others though, such as the Daily Mail article *How comeback kid Sweden got the last laugh on coronavirus* [40] from the same day was more positive. The Daily Mail argued that, Sweden's infection rate was now lower than in UK, Spain, France, Italy or Denmark, that the curve was flattened without a lockdown, and that Swedish economy had seen a milder downturn than in much of Europe.

So, at this point the infection level had gone down, and compared to other European countries it seemed to head in the right direction. But of course, there was still a long path to go before the pandemic and its long-term impact on economy, education, and health, could be summarized.

4   Analysis from January 28, 2021

Just when I ended my last analysis, on September 15, the Public Health Authority recommended to lift the curfew in elderly care [41]. According to the Director General Johan Carlson, the recommendation was motivated by the risk for negative consequences for the physical and mental health of the elderly, the increased knowledge about preventive measures and the fact that the spread of infection had decreased sharply.

At this point coronavirus cases were rising in pretty much all other European countries; however, it was looking pretty good in Sweden. Anders Tegnell commented that Sweden was expected to have *"a low level of spread with occasional local outbreaks"*, as covered in the article *Anders Tegnell and the Swedish Covid experiment* in the Financial Times on September 11 [42].

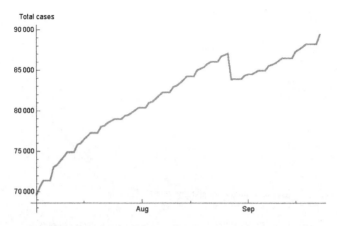

Figure 16: Accumulated number of known cases with COVID-19 in Sweden from June 31 to September 22 2020.

Only eleven days later, on September 22 [43], the Public Health Authority stated that they were seeing an increase of COVID-19 cases in several regions, in all age groups except the elderly. They also mention that a lot of the cases were related to sports, and especially football (soccer) and ice hockey. As Fig. 16 shows, this increasing trend was not easy to spot by merely looking at the number of known cases. The sudden decrease on August 27 is explained by a correction that was made after finding out that tests used in nine different regions had been giving false positives [46]. On September 29, a decision to lift the curfew in elderly care from October 1 was taken [44].

Comparing regions

To this point, the same recommendations had been applied to the entire country. However, on October 13 a decision is taken to allow for local recommendations from October 19 [45]. These local recommendations should always be time limited to three weeks, but with the possibility to be extended. The day after, October 20 (Tuesday of week 43), the region of Uppsala added two local recommendations [47]:

- Avoid travelling by public transport or other public transport.

- If possible, avoid having physical contact with people other than those you live with.

Fig. 17 shows the number of confirmed daily cases per 100,000 in each Swedish region, according to data downloaded from the Public Health Authorities on January 28, 2021. At this time, the cases in Uppsala were on

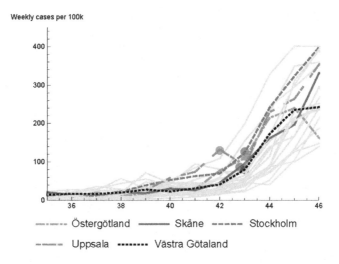

Figure 17: Weekly number of known cases per 100,000 population in the different Swedish regions, highlighting Östergötland, Skåne, Stockholm, Uppsala, and Västra Götaland. The circles indicate the point at which temporary local recommendations were added in those five regions.

the rise, with more cases per 100,000 than any other Swedish region in week 42. The blue dot indicates when the local recommendation was introduced.

The region of Skåne decided to introduce local recommendations on October 27 [48], then Stockholm, Västra Götaland, and Östergötland all did the same on October 29 [49].

During the following weeks Jönköping, Halland, Örebro [50], Kronoberg and Södermanland [51], Kalmar, Västerbotten, and Norrbotten [52], Dalarna, Gotland, Värmland, and Västmanland [53], Gävleborg and Västernorrland [54], Jämtland [55], and Blekinge [56] all followed suit. Some of these were extended after their initial three weeks period, but from December 14 no region was applying local recommendations.

At the same time, during November and December, several adjustments were made in national recommendations. These included reducing the maximum group size in restaurants and bars to 8 [57], and a recommendation for high schools to go back to distance teaching [58]. On December 18 several new recommendations were added [59]. These included, further reducing maximum size of groups at restaurant to 4 persons, advising businesses to cancel sales during the holidays, and announcing that a recommendation to wear masks when using public transport at peak hours traffic would be added from January 7 (when most Swedes would be back to work from the holidays).

November 26, the Public Health Authority stated that the spread of COVID-19 could reach its peak in mid-December [60], which when looking

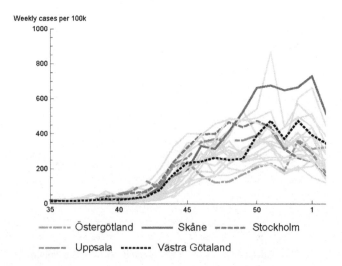

Figure 18: Weekly number of known cases per 100,000 persons in the different Swedish regions, highlighting Östergötland, Skåne, Stockholm, Uppsala, and Västra Götaland.

at the number of cases per region, see Fig. 18 turned out to be fairly accurate. However, while this prediction was accurate, the prediction from the beginning of September ("*a low level of spread with occasional local outbreaks*"), was proven wrong.

Let us compare the number of accumulated deaths with COVID-19 in different Swedish regions over the last year, Fig. 19. As mentioned, the strategy has been the same throughout the country for almost the whole time, except some minor differences especially between mid-October to mid-December. Despite this, some regions have feared relatively well throughout the whole period, others did well in the first wave but not the second, while some were hit harder in both waves as shown in Fig. 21. Thus, despite the same strategy, the result varies substantially.

Comparing with other countries

Now, let us look how Sweden compares with other countries. Fig. 20, shows accumulated number of deaths per 100,000 in Sweden, together with the other countries in the European Union, the G7 countries, and Sweden's closest neighbours. Accumulated deaths are more reliable than number of cases, however registration will differ between countries as previously mentioned. That said, for the European Union, G7, and Sweden's neighbours, it seems to be relatively accurate and comparable. Still, the exact numbers should be taken with a bit of care.

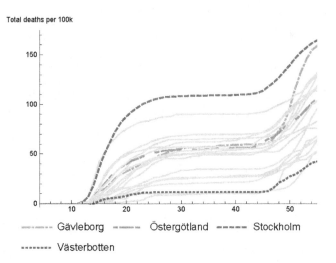

Figure 19: Accumulated number of deaths with COVID-19 per 100,000 population in the different Swedish regions, highlighting Gävleborg, Östergötland, Stockholm, and Västerbotten.

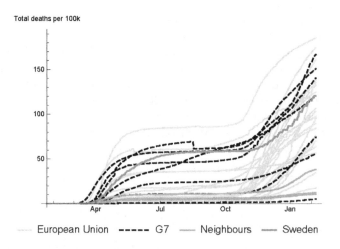

Figure 20: Total number of confirmed deaths with COVID-19 in Sweden compared with the different countries in the European Union, G7, and Sweden's three closest neighbours (Denmark, Finland, and Norway).

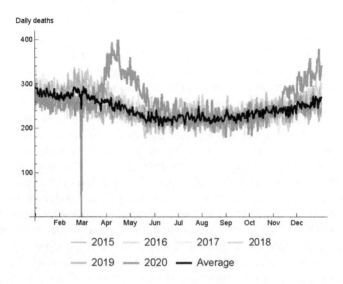

Figure 21: Daily number deaths in Sweden from 2015 to 2020.

It is clear that most countries got a second (one can argue tat for a few it was their first). When comparing with other EU and G7 countries, Sweden has a similar trajectory, despite a different strategy. The result was worse then most others during the first wave, but possibly slightly better in the second. However, comparing with the closest neighbours (with a strategy somewhere in between Sweden and the most of the other countries in the comparison) it is clear that they have managed to keep death tolls substantially lower. Yet again the opinion in media shifts a bit, as illustrated i the article *Sweden's second wave offers hard reality check* (Japan Times, December 19) [61].

Finally, let's see a comparison with previous years again. Looking at Fig. 21 it is easy to spot both wave one and two.

Additional notes

Some final notes are in place. Sweden started vaccination on December 27, focusing on people at elderly care [62]. On January 8, 2021, a new legislation, temporarily giving the government the power to decide in an ordinance on more binding infection control measures than was previously possible. The possibility was used immediately when some of the previous recommendations for businesses was changed to binding regulations [63]. The changes made were very limited and probably did not affect too much.

5   Conclusion

Using basic data analysis in the Wolfram Language the spread of COVID-19 in Sweden has been analysed at four different stages of the pandemic. The decisions taken by the Public Health Authority of Sweden has been set in relationship to the progress of the pandemic. Despite that all Swedish regions have had very similar restrictions the difference between Swedish regions is relatively large. In fact, it is comparable to the difference between different European and G7 countries, which all have had very different strategies to Sweden. However, with strategies that have been somewhere in between Sweden's and the rest of the European Union, Sweden's neighbours have had substantially lower death rates than Sweden, as well as the rest of Europe. Thus, at this point there is no clear indication of *the Swedish Experiment* being either better or worse than other strategies when it comes to stopping the infection. In fact, the spread between different Swedish regions is fascinatingly similar to the spread between EU and G7 countries.

While many try to describe things in black and white, my conclusion from looking at data is that, at this moment, the picture is very much painted in grey. Yet again, the pandemic is far from over, and to make any real verdict on the success of any strategy secondary effects, such as the effect on economy, employment, and physical and mental health, has to be accounted for.

Bibliography

[1] Wolfram Language, https://www.wolfram.com/language/

[2] COVID-19 - the Swedish experiment - is it working?, Jan Brugård, https://community.wolfram.com/groups/-/m/t/1974412

[3] Bekräftat fall i Jönköping av nya coronaviruset (2019-nCoV), Public Health Agency of Sweden, https://www.folkhalsomyndigheten.se/nyheter-och-press/nyhetsarkiv/2020/januari/bekraftat-fall-av-nytt-coronavirus-i-sverige/

[4] Information om karantän, Public Health Agency of Sweden, https://www.folkhalsomyndigheten.se/nyheter-och-press/nyhetsarkiv/2020/februari/information-om-karantan/

[5] Förändrad riskbedömning för fall av covid-19 i Sverige, Public Health Agency of Sweden, https://www.folkhalsomyndigheten.se/nyheter-och-press/nyhetsarkiv/2020/februari/forandrad-riskbedomning-for-fall-av-covid-19-i-sverige/

[6] Nytt bekräftat fall av covid-19, Public Health Agency of Sweden, https://www.folkhalsomyndigheten.se/nyheter-och-press/nyhetsarkiv/2020/februari/nytt-bekraftat-fall-av-covid-19/

[7] Ytterligare fall av covid-19 i flera regioner, Public Health Agency of Sweden, https://www.folkhalsomyndigheten.se/nyheter-och-press/nyhetsarkiv/2020/februari/ytterligare-fall-av-covid-19-i-flera-regioner/

[8] Pressinbjudan - Folkhälsomyndigheten begär att trafiktillståndet för IranAir omprövas, Public Health Agency of Sweden, https://www.folkhalsomyndigheten.se/nyheter-och-press/nyhetsarkiv/2020/mars/pressinbjudan---folkhalsomyndigheten-begar-att\-trafiktillstandet-for-iranair-omprovas/

[9] Uppdaterad riskbedömning för covid-19 i Sverige, Public Health Agency of Sweden, https://www.folkhalsomyndigheten.se/nyheter-och-press/nyhetsarkiv/2020/mars/uppdaterad-riskbedomning-for-covid-19-i-sverige/

[10] ökad spridning av covid-19 i norra Italien, Public Health Agency of Sweden, https://www.folkhalsomyndigheten.se/nyheter-och-press/nyhetsarkiv/2020/mars/okad-spridning-av-covid-19-i-norra-italien/

[11] Folkhälsomyndigheten har rekommenderat avrådan från resor till norra Italien, Public Health Agency of Sweden, https://www.folkhalsomyndigheten.se/nyheter-och-press/nyhetsarkiv/2020/mars/folkhalsomyndigheten-har-rekommenderat-avradan-fran-resor\-till-norra-italien/

[12] Flera tecken på samhällsspridning av covid-19 i Sverige, Public Health Agency of Sweden, https://www.folkhalsomyndigheten.se/nyheter-och-press/nyhetsarkiv/2020/mars/flera-tecken-pa-samhallsspridning-av-covid-19-i-sverige/

[13] Förslag: Inga allmänna sammankomster med fler än 500 personer, Public Health Agency of Sweden, https://www.folkhalsomyndigheten.se/nyheter-och-press/nyhetsarkiv/2020/mars/forslag-inga-allmanna-sammankomster-med-fler-an-500-personer/

[14] Spridningen av covid-19 är en pandemi, Public Health Agency of Sweden, https://www.folkhalsomyndigheten.se/nyheter-och-press/nyhetsarkiv/2020/mars/spridningen-av-covid-19-ar-en-pandemi/

[15] Ny fas kräver nya insatser mot covid-19, Public Health Agency of Sweden, https://www.folkhalsomyndigheten.se/nyheter-och-press/nyhetsarkiv/2020/mars/ny-fas-kraver-nya-insatser-mot-covid-19/

[16] Personer över 70 bör begränsa sociala kontakter tills vidare, Public Health Agency of Sweden, https://www.folkhalsomyndigheten.se/nyheter-och-press/nyhetsarkiv/2020/mars/personer-over-70-bor-begransa-sociala-kontakter-tills-vidare/

[17] Lärosäten och gymnasieskolor uppmanas nu att bedriva distansundervisning, Public Health Agency of Sweden, https://www.folkhalsomyndigheten.se/nyheter-och-press/nyhetsarkiv/2020/

mars/larosaten-och-gymnasieskolor-uppmanas-nu-att-bedriva\
-distansundervisning/

[18] Tänk över om resan verkligen är nödvändig, Public Health Agency
of Sweden, https://www.folkhalsomyndigheten.se/nyheter-och-press/
nyhetsarkiv/2020/mars/tank-over-om-resan-verkligen-ar-nodvandig/

[19] Nya regler för restauranger och krogar, Public Health Agency
of Sweden, https://www.folkhalsomyndigheten.se/nyheter-och-press/
nyhetsarkiv/2020/mars/nya-regler-for-restauranger-och-krogar/

[20] Förslag: Ytterligare   begränsningar   av   allmänna   sammankomster,   Public   Health   Agency   of   Sweden,   https://www.
folkhalsomyndigheten.se/nyheter-och-press/nyhetsarkiv/2020/mars/
forslag-ytterligare-begransningar-av-allmanna-sammankomster/

[21] Sweden bucks global trend with experimental virus strategy, Financial Times,
https://www.ft.com/content/31de03b8-6dbc-11ea-89df-41bea055720b

[22] In the Coronavirus Fight in Scandinavia, Sweden Stands Apart,
NY   Time,   https://www.nytimes.com/2020/03/28/world/europe/
sweden-coronavirus.html

[23] Coronavirus, la Svezia va controcorrente: mezzi di trasporto pieni e uffici
aperti, La Repubblica, https://www.repubblica.it/esteri/2020/03/26/
news/coronavirus_la_svezia_va_controcorrente_mezzi_di_trasporto_
pieni_e_uffici_aperti-252380584/

[24] La excepción sueca, El País, https://elpais.com/sociedad/2020-04-01/
la-excepcion-sueca.html

[25] Updates during April, Public Health Agency of Sweden, https://www.
folkhalsomyndigheten.se/nyheter-och-press/nyhetsarkiv/2020/april/

[26] 412,000 Missing Deaths: Tracking the True Toll of the Coronavirus
Outbreak, NY Times, https://www.nytimes.com/interactive/2020/04/21/
world/coronavirus-missing-deaths.html

[27] överdödligheten   sjunker   i   Sverige,   SCB   (Statistics   Sweden),
https://www.scb.se/om-scb/nyheter-och-pressmeddelanden/
overdodligheten-sjunker-i-sverige/

[28] Preliminary statistics over deaths (continuously updated), SCB (Statistics
Sweden), https://www.scb.se/hitta-statistik/statistik-efter-amne/
befolkning/befolkningens-sammansattning/befolkningsstatistik/
pong/tabell-och-diagram/preliminar-statistik-over-doda/

[29] Första resultaten från pågående undersökning av antikroppar för
covid-19-virus, Public Health Agency of Sweden, https://www.
folkhalsomyndigheten.se/nyheter-och-press/nyhetsarkiv/2020/maj/
forsta-resultaten-fran-pagaende-undersokning-av-antikroppar\
-for-covid-19-virus/

[30] Idrottstävlingar och matcher tillåts, Public Health Agency of Sweden, https://www.folkhalsomyndigheten.se/nyheter-och-press/nyhetsarkiv/2020/maj/idrottstavlingar-och-matcher-tillats/

[31] Befolkningsutvecklingen i riket efter kön. år 1749 - 2019, SCB (Statistics Sweden), http://www.statistikdatabasen.scb.se/pxweb/sv/ssd/START$_ _$BE$__$BE0101$__$BE0101G/BefUtvKon1749/

[32] Goda exempel för att minska smittspridning på äldreboenden, Public Health Agency of Sweden, https://www.folkhalsomyndigheten.se/nyheter-och-press/nyhetsarkiv/2020/juni/goda-exempel-for-att-minska-smittspridning-pa-aldreboenden/

[33] Reviderat stöd för provtagning för covid-19 inom äldreomsorgen, Public Health Agency of Sweden, https://www.folkhalsomyndigheten.se/nyheter-och-press/nyhetsarkiv/2020/juni/reviderat-stod-for-provtagning-for-covid-19-inom\-aldreomsorgen/

[34] Ny restauranglag ska bidra till minskad spridning av covid-19, Public Health Agency of Sweden, https://www.folkhalsomyndigheten.se/nyheter-och-press/nyhetsarkiv/2020/juli/ny-restauranglag-ska-bidra-till-minskad-spridning-av\-covid-19/

[35] Nya föreskrifter och allmänna råd till serveringsställen, Public Health Agency of Sweden, https://www.folkhalsomyndigheten.se/nyheter-och-press/nyhetsarkiv/2020/juli/nya-foreskrifter-och-allmanna-rad-till-serveringsstallen/

[36] Fortsätt arbeta hemma om det finns möjlighet, Public Health Agency of Sweden, https://www.folkhalsomyndigheten.se/nyheter-och-press/nyhetsarkiv/2020/juli/fortsatt-arbeta-hemma-om-det-finns-mojlighet/

[37] Körer får rekommendationer om mer smittsäker verksamhet, Public Health Agency of Sweden, https://www.folkhalsomyndigheten.se/nyheter-och-press/nyhetsarkiv/2020/augusti/korer-far-rekommendationer-om-mer-smittsaker-verksamhet/

[38] Barn och unga med symtom på covid-19 rekommenderas PCR-testning för snabbare återgång till skolan, Public Health Agency of Sweden, https://www.folkhalsomyndigheten.se/nyheter-och-press/nyhetsarkiv/2020/augusti/barn-och-unga-med-symtom-pa-covid-19-rekommenderas-pcr\-testning-for-snabbare-atergang-till-skolan/

[39] Anders Tegnell and the Swedish Covid experiment, Financial Times, https://www.ft.com/content/5cc92d45-fbdb-43b7-9c66-26501693a371

[40] How comeback kid Sweden got the last laugh on coronavirus, Daily Mail, https://www.dailymail.co.uk/news/article-8722051/How-comeback-kid-Sweden-got-laugh-coronavirus.html

[41] Förslag om att besöksförbudet på särskilda boenden ska upphöra, Public Health Agency of Sweden, https://www.folkhalsomyndigheten.se/nyheter-och-press/nyhetsarkiv/2020/september/forslag-om-att-besoksforbudet-pa-sarskilda-boenden-ska-upphora/

[42] Anders Tegnell and the Swedish Covid experiment, Financial Times, https://www.ft.com/content/5cc92d45-fbdb-43b7-9c66-26501693a371

[43] Flera utbrott av covid-19 inom idrottslag och i samband med tävlingar, September 22, 2020 - https://www.folkhalsomyndigheten.se/nyheter-och-press/nyhetsarkiv/2020/september/flera-utbrott-av-covid-19-inom-idrottslag-och-i-samband\
-med-tavlingar/

[44] Alla har ett stort ansvar för att minimera smittspridning när besöksförbudet upphör, https://www.folkhalsomyndigheten.se/nyheter-och-press/nyhetsarkiv/2020/september/alla-har-ett-stort-ansvar-for-att-minimera-smittspridning\
-nar-besoksforbudet-upphor/

[45] Lokala allmänna råd vid utbrott av covid-19, https://www.folkhalsomyndigheten.se/nyheter-och-press/nyhetsarkiv/2020/oktober/lokala-allmanna-rad-vid-utbrott-av-covid-19/

[46] Brister i coronatest gav tusentals falska svar, https://www.lakemedelsvarlden.se/brister-i-coronatest-gav-3-700-falska-svar/

[47] Beslut om skärpta allmänna råd i Uppsala län, https://www.folkhalsomyndigheten.se/nyheter-och-press/nyhetsarkiv/2020/oktober/beslut-om-skarpta-allmanna-rad-i-uppsala-lan/

[48] Beslut om skärpta allmänna råd i Skåne län, https://www.folkhalsomyndigheten.se/nyheter-och-press/nyhetsarkiv/2020/oktober/beslut-om-skarpta-allmanna-rad-i-skane-lan/

[49] Beslut om skärpta allmänna råd i Stockholms län, Västra Götalands län och Östergötlands län, https://www.folkhalsomyndigheten.se/nyheter-och-press/nyhetsarkiv/2020/oktober/beslut-om-skarpta-allmanna-rad-i-stockholms-lan-vastra\
-gotalands-lan-och-ostergotlands-lan/

[50] Beslut om skärpta allmänna råd i Jönköpings län, Hallands län och Örebro län, https://www.folkhalsomyndigheten.se/nyheter-och-press/nyhetsarkiv/2020/november/beslut-om-skarpta-allmanna-rad-i-jonkopings-lan-hallands\
-lan-och-orebro-lan/

[51] Beslut om skärpta allmänna råd i Kronobergs och Södermanlands län, https://www.folkhalsomyndigheten.se/nyheter-och-press/nyhetsarkiv/2020/november/beslut-om-skarpta-allmanna-rad-i-kronobergs-och-sormlands-lan/

[52] Beslut om skärpta allmänna råd i Kalmar-, Norrbottens- och Västerbottens län, `https://www.folkhalsomyndigheten.` `se/nyheter-och-press/nyhetsarkiv/2020/november/` `beslut-om-skarpta-allmanna-rad-i-kalmars--norrbottens\` `--och-vasterbottens-lan/`

[53] Beslut om skärpta allmänna råd i Dalarnas-, Gotlands-, Värmlands och Västmanlands län, `https://www.` `folkhalsomyndigheten.se/nyheter-och-press/nyhetsarkiv/2020/` `november/beslut-om-skarpta-allmanna-rad-i-dalarnas--gotlands\` `--varmlands-och-vastmanlands-lan/`

[54] Beslut om skärpta allmänna råd i Gävleborgs- och Västernorrlands län, `https:` `//www.folkhalsomyndigheten.se/nyheter-och-press/nyhetsarkiv/` `2020/november/beslut-om-skarpta-allmanna-rad-i-gavleborgs--och-\` `vasternorrlands-lan/`

[55] Beslut om skärpta allmänna råd i Jämtlands län, `https://www.` `folkhalsomyndigheten.se/nyheter-och-press/nyhetsarkiv/2020/` `november/beslut-om-skarpta-allmanna-rad-i-jamtlands-lan/`

[56] Beslut om skärpta allmänna råd i Blekinge län, `https://www.` `folkhalsomyndigheten.se/nyheter-och-press/nyhetsarkiv/2020/` `november/beslut-om-skarpta-allmanna-rad-i-blekinge-lan/`

[57] Nya föreskrifter och allmänna råd till serveringsställen, `https://www.folkhalsomyndigheten.` `se/nyheter-och-press/nyhetsarkiv/2020/november/` `nya-foreskrifter-och-allmanna-rad-till-serveringsstallen/`

[58] Folkhälsomyndigheten rekommenderar att gymnasieskolorna övergår till distansundervisning, `https://www.folkhalsomyndigheten.` `se/nyheter-och-press/nyhetsarkiv/2020/december/` `folkhalsomyndigheten-rekommenderar-att-gymnasieskolorna\` `-overgar-till-distansundervisning/`

[59] Ytterligare åtgärder för att bromsa smitta och skydda hälsa, `https://www.folkhalsomyndigheten.` `se/nyheter-och-press/nyhetsarkiv/2020/december/` `ytterligare-atgarder-for-att-bromsa-smitta-och-skydda-halsa/`

[60] Smittspridningen kan nå kulmen i mitten av december, `https:` `//www.folkhalsomyndigheten.se/nyheter-och-press/nyhetsarkiv/2020/` `november/smittspridningen-kan-na-kulmen-i-mitten-av-december/`

[61] Smittspridningen kan nå kulmen i mitten av december, `https:` `//www.folkhalsomyndigheten.se/nyheter-och-press/nyhetsarkiv/2020/` `november/smittspridningen-kan-na-kulmen-i-mitten-av-december/`

[62] Sweden's second wave offers hard reality check, `https://www.` `japantimes.co.jp/opinion/2020/12/19/commentary/world-commentary/` `sweden-coronavirus-lessons/`

[63] Butiker, gym och köpcentrum måste begränsa antalet besökare, https://www.folkhalsomyndigheten.se/nyheter-och-press/nyhetsarkiv/2021/januari/butiker-gym-och-kopcentrum-maste-begransa-antalet-besokare/

Procurement Strategies for Personal Protective Equipment: Insuring the Healthcare System Against Pandemic

Matthew J DiPaola, MD*, Christian P DiPaola, MD and David Salazar

*Corresponding author: mdipaola@buffalo.edu

Pandemics put health care systems under extreme stress due to high demand surges for health care services. As demand for patient care services spike, so to does the demand for personal protective equipment (PPE). Personal protective equipment is one major means by which a health system insures its staff and patients against disease and thus slow pandemic spread. Health care institutions have been caught unprepared for the volume of PPE necessary to properly protect the health care workforce during the COVID 19 pandemic. Health care systems must adopt PPE procurement and readiness strategies that respect the unpredictable and rapidly evolving dynamics of pandemics. In this paper we discuss principles of PPE procurement that should be considered when preparing any health system for future pandemic. We discuss the need for health systems to consider a shift toward locally produced and/or reusable PPE to prevent shortages in future pandemic situations.

1 Introduction

As the COVID19 pandemic unfolded, drastic shortages of critical supplies of personal protective equipment (PPE) arose nearly ubiquitously across the globe, exposing a lethal vulnerability in our current health care framework [1]. Stories of health care workers in New York City substituting garbage bags for standard issue PPE blanketed the news early in the crisis, illustrating the dire shortages with which health systems in even developed nations struggled [2, 3]. Under current procurement arrangements, the evidence illustrates that pandemics produce severe resource shortages. Current PPE supply models appear to be heavily reliant on fluidly moving global supply chains that have failed to function properly during crisis.

These shortages have led leadership bodies such as the CDC to sanction PPE rationing strategies to conserve equipment. For example the CDC lifted strict use criteria of N95 respirator masks and advocated for sterilization and reuse beyond the typical 4-hour rating window [4]. Others recommended that governments monitor national face mask supplies and regulate their usage recommendations accordingly [5]. Maintaining adequate stocks of PPE is vital. Failure to protect the health care work force during a pandemic can create a negative feedback loop where the health care capacity as a function of provider availability diminishes while demand for services surge [1,3,6].

Some have attempted to model epidemiological phenomena as a means to better forecast the extent of infectious disease outbreaks and influence resource allocation to meet PPE demand [1,7,8]. For the most part however, these attempts have resulted inadequate. Gooding [10] points out that while "there exist wide bodies of literature in epidemiology and in emergency response logistics, there is remarkably little research on the connection between the two," contending that demand forecasting assumptions are often too simplistic and assume that supply chains will be perfectly reactive during crisis [10]. The author surveyed many front-line procurement agents of the Ebola pandemic and found that even though some were able to use epidemiological models to help predict PPE demand usage, many noted that the models were often overly complex and failed for non epidemiological reasons as well [10]. For instance procurement agents were often locked into certain distributor contracts and health care workers often made "conservative" decisions choosing to "over" protect due to lack of trust of some PPE options throwing off prediction models [10].

Others have examined "pre-positioning" as a means of preparing stocks of PPE for disaster situations. "Pre positioning" is a term used to describe allocating caches of resources to "optimal" locations near potential disaster zones prior to emergency events to account for transportation and demand uncertainty [9,11]. Pre-positioning strategies have been criticized for being overly expensive if disaster fails to hit in the "optimal" location. But what if one cannot predict the "optimal" location? Pandemic is a unique type of disaster that occurs everywhere at once. Pre-positioning may fail simply because there is no "optimal" location if a disaster is "everywhere." A broader strategy that respects this dynamic will be necessary to prevent future PPE procurement issues.

The failure of these previous strategies likely stems from a failure to implement sound risk management practices that account for the inherent unpredictability of the next pandemic. Indeed, the previous strategies rely on over-optimized predictions for the timing and scope of the next global emergency. Therefore, the success of any PPE procurement strategy becomes extremely fragile to the success of these predictions. However, as Taleb and Cirillo [12] have shown, it is fundamentally impossible to accurately predict the size and duration of the next pandemic. Indeed, these variables belong to the class of the most difficult variables to predict: fat-tailed variables [13].

Briefly a "fat tailed" probability distribution is one where all its statistical properties are dominated by a few observations [13]. Mathematically, a variable is fat-tailed if its survival function is regularly varying: $S(x) = L(x)x^{1/\xi}$, where $L(x)$ is a slowly varying function such that $\lim_{x \to \infty} \frac{L(cx)}{L(x)} = 1$ for $c > 0$ [14]. Where ξ is the tail parameter that controls how slowly the survival function decreases. Therefore, for small values of ξ, even as the events under study get closer and closer to the extreme, they are still plausible and they define any of the statistical moments of the distribution.

Given the profound importance of the extreme events, the study of the mathematical properties and real life implications of such variables is the study of their extremes: their maxima and minima. Extreme Value Theory (EVT) is a field of statistics that deals with these types of rare but impactful events. Therefore, we argue that EVT may offer a better guidepost for developing PPE procurement and distribution strategies for healthcare systems than current methods.

In this paper we will address how health care systems should alter their PPE procurement strategies to account for the risk of future pandemic. We will explore PPE procurement as a function of a balance between disposable vs reusable stock and how systems over optimized toward disposable PPE may become fragile to supply shortages during pandemics. We will explain why health care systems should reanalyze their PPE strategy and consider a shift toward more reusable and locally sourced stocks of PPE in order to insure their health care work force against future catastrophic pandemic events.

2 Current Health care System: Disposable PPE

Disposable PPE has become standard in the US health system in recent times. Disposable PPE confers many advantages to the user, the patient and the healthcare system. It is both convenient and safe for the end users. It allows rapid use and changeover between patients with minimal need for user attention to equipment function and cleanliness prior to usage. Once it is used it can be disposed of decreasing risk of cross contamination. Disposability reduces complexity of storage and shipping. All of these benefits are desirable. However they all come at a cost, a potentially dangerous cost if the system over-optimizes too heavily toward this method alone. The cost to the system is that the strategy may become too fragile to unpredictable massive demand shocks. We examine the scenario of a system that is reliant on disposable PPE.

Once committed almost solely to a system of disposable PPE, health care facilities depend on a constant influx of new stock to replenish supply. For a health system in a steady equilibrium state, replenishing disposable PPE is a matter of routine: estimate monthly usage and leave a reasonable cushion for normal fluctuations around the mean. Pandemics however, upend mean-based estimates leading to shortages and system failure. Pandemics inflict a double insult to the system: 1) a rapid spike in demand for PPE and

2) frozen resource supply chains. When this happens stocks of disposable PPE become unavailable and healthcare facilities often adopt alternative strategies such as rationing which may put healthcare workers at risk. Where "just-in-time" supply policies may suffice in conventional times, they assuredly fall flat in the midst of pandemic.

3   Pandemics are Unpredictable

Evidence from Cirillo and Taleb, 2020

Taleb and Cirillo [12] have shown that pandemics are patently fat-tailed phenomena through a rigorous study of historical records for the last 72 major pandemic diseases. Mathematically they define the distribution of casualties as $G(z)$. Thanks to Pickands–Balkema–de Haan theorem [14], they can approximate the distribution of tail events with a Generalized Pareto Distribution (GPD) thus:

$$G_u(z) = P(Z \leq z \mid Z > u) = \frac{G(z) - G(u)}{1 - G(u)} \approx GPD(z; \xi, \beta, u)$$

Using maximum likelihood, Taleb and Cirillo [12] estimate a tail parameter around $\xi = 1.62$ (*se* = 0.52). Therefore, the authors conclude that pandemic fatalities are "an extremely erratic phenomenon, with substantial tail risk". Indeed, the result clearly rejects the possibility of a second finite moment, thus rendering the use of any prediction method that relies on the sample mean "too volatile to be safely used" [12].

From a lack of scale to unpredictability

With this context, we can now better understand the enormous challenge that we face when we try to predict the next pandemic. The casualties, the duration, the number of infected people at any point in time, are functions of a phenomena where the uncertainty is maximal. This unpredictability arises from the lack of a characteristic scale for fat-tailed variables. Mathematically (Embrechts, 1997), we can express this lack of characteristc scale thus:

$$\lim_{K \to \infty} \frac{1}{K} E(X|X > K) = \lambda, \lambda > 1$$

Intuitively, as Taleb [13] states in Statistical Consequences of Fat Tails: "There is no typical collapse or disaster, owing to the absence of a characteristic scale". That is, there is no "typical" sized flood, hurricane, or pandemic.

Therefore, pandemic processes can deliver an enormous range of devastating effects with no predictable ceiling, as the uncertainty around the extremes is so large. Additionally the size and scope of past pandemics may bear no familiar resemblance to the size and scope of future pandemics: they may be much more mild or devastatingly more severe. It follows then that PPE demand may in turn be unpredictable. Both the amount and longevity of usage of PPE in future pandemics will vary as a function of the next pandemic's longevity scale and scope, two variables that we cannot predict with any amount of precision. Not even taking COVID yearly levels of PPE as our baseline.

Not even the past can guide us

Indeed, suppose that you take our current (mid pandemic) yearly PPE needs as the stockpile that will be held for the future. This is not wise risk management. It incurs in the "Lucretius fallacy": mistaking the largest flood ever seen for the largest flood that could be seen. Just as previous historical records for PPE stockpiles resulted ineffective for the current pandemic, they surely will result ineffective for the next pandemic that will surpass the current one.

Imagine a future pandemic whose R_0 is twice as great because it has multiple mechanisms of infectivity. Predicted PPE usage even akin to COVID usage standards still may not account for the demand spike in such a case.

Mathematically, we can explain this phenomenon thus. If X is fat-tailed, it is also long-tailed. Therefore, the probabilities of an extreme and another, even larger extreme are connected thus:

$$\lim_{x \to \infty} \Pr[X > x + t \mid X > x] = 1$$

If the probability of our current extreme, COVID, is not negligible, then neither it is the probability of an even worse pandemic. Therefore, there's no amount of disposable PPE that can protect us for the next potential pandemic.

No risk minimization, but risk management

We can model mathematically this fragility to our predictions by posing a payoff as a function of our estimated needs for PPE $\hat{y}$ and the realized needs, y:

$$g(y, \hat{y}) = min(0, \hat{y} - y)$$

We can deal with the uncertainty around this payoff in two ways.

First, we can claim that we shouldn't over-react: after all, the chance of any pandemic hitting us in a particular year is very small. Mathematically, we can focus on minimizing the following function: the payoff at K (a "large" stockpile of PPE) multiplied by the probability of our needs exceeding K:

$$I_2 = g(K) \int_K^\infty f(y)dy = g(K)p_K$$

However, due to the lack of a characteristic scale in the pandemic casualties and infected people, we can convince ourselves that there's no large enough K for the above observation to be meaningful. Secondly, then, we can realize that probability of needing more than K is never negligible, and neither is the very real negative payoff that we would face if we don't have enough PPE to protect ourselves.

Indeed, Taleb [13] shows that, under fat-tails, the real integral to calculate the random payoff of just stockpiling K is the following:

$$I_1 = \int_K^\infty g(x)f(x)\mathrm{d}x$$

There's no bound to how much worse we can do. Not even taking disposable PPE we can stockpile, we can still under-predict the size of the next pandemic and still suffer from a critical shortage of PPE and have a very, very negative payoff. The equivalent of the COVID pandemic that we are still experiencing. The failure to understand this critical point between I_1, I_2 is called the "ludic fallacy" and has been exposed by Taleb extensively in his book The Black Swan [15].

At this point, we can draw many similarities from the world of finance. Only stockpiling K amount of disposable PPE would be equivalent to trying to hedge a continuous exposure with a binary bet. In Taleb's words, with yet another parallel to health insurance: "equivalent to a health insurance payout of a lump sum if one is "very ill"– regardless of the nature and gravity of the illness" [13]

Therefore, a new standard for PPE sourcing is needed. Following finance and health insurance, we know that we cannot rely on risk minimization but risk management. Indeed, sound risk management is "about changing the payoff function g(.) rather than making "good forecasts" [13]. In our case, this means that we cannot rely on accurately predicting how much disposable PPE we will need. We need to find a way to fulfill our PPE needs regardless of the size of the next global pandemic.

4   A New Standard

There is one variable that is relatively predictable at any point in time for a given healthcare system: the number of healthcare workers. We should thus direct our efforts at changing the payoff function as opposed to attempting to predict the size of the next pandemic. Treat PPE like insurance for your workforce. At any one time the number of healthcare workers in a system is relatively constant (the United States has roughly 17 million healthcare workers). Since this number is relatively constant one could shift one's perspective and plan for a specific dollar amount that it would take to maximally protect all 17 million members of a health system's work force at a given time. We believe that a shift toward greater use of highly protective, reusable PPE offers at least a partial solution to this problem.

5   The Outlier

Health care workers are at risk for COVID. As of January 29, 2021 the CDC had reported over 385,000 confirmed COVID cases among U.S. health care workers (a number that likely widely underestimates the problem as only about 19 million people were sampled for the data [4]. Others have shown that COVID infection rates of the health care workforce mirrors that of the general population with some specialties being particularly susceptible due to high workforce exposure [16]. There has been at least one notable exception.

In April 2020 during the height of the pandemic Cotugno Hospital (Naples, Italy), a hospital that was filled to capacity with COVID cases, had reported an astonishing ZERO COVID cases among its workforce [17]. Of note Cotugno hospital is a specialty infectious disease hospital and employs much more stringent PPE guidelines than the typical health system. Cotugno assumes that all of their patients are carriers of highly infectious diseases. Their PPE protocols are geared toward full body protection of their workforce most often with reusable suits [17]. This outlier may provide clues for how other health systems could protect their workforce in future pandemics.

6   The Current System in Most Health care Facilities

In its current form, the majority of other health care facilities around the globe rely on PPE strategies that rely on "just in time" supply chains that stock predominantly disposable PPE. Unfortunately as we have seen during this pandemic supply chains often choke and shortages of PPE and other materials result. This calls into question both the limits of a "just in time" philosophy for procuring PPE but the limits of a global supply chain.

7   Reusable Personal Protective Equipment

In contrast to disposable PPE, reusable PPE has a different set of trade-offs. Reusable PPE requires the user to ensure cleanliness before usage. It is typically more expensive upfront. It also requires space for storage and staff must ensure consistent cleaning and maintenance protocols. But reusable PPE also has a unique advantage: supply of most reusable PPE does not dwindle significantly in times of supply-chain disruption. Stated succinctly: as long it is well maintained, reusable PPE offers the advantage of durability and availability in times of high system stress and demand. Some have shown that reusable PPE options such as isolation gowns can be cost effective when compared with disposable alternatives [18] [19].

Health care systems by their nature must function fluidly in both steady state dynamics in which PPE usage runs on a more predictable schedule

PPE Type	Advantages	Disadvantages
Disposable	-Safe: low risk of cross contamination -Allows for rapid change of equipment for personnel -Minimal need for maintenance -Comes sterile packaged	-Requires hospitals to continually replenish stock -Stocks are susceptible to resource squeezes and supply chain disruption
Reusable	-More robust to supply chain disruptions in crisis situation (assuming systems stay well stocked)	-Requires health facilities to have storage, maintenance and sterilization protocols -Often higher upfront cost

as well as surge states when the system sees a drastic increase in resource utilization as occurs in pandemic scenarios. Pandemics have the potential to cripple the health care delivery system through contagion of the health care workforce. If enough workers go out sick as demand for services increases, the system buckles and fails in its mission to care for the population.

In light of the above mission and the stated advantages and disadvantages of both disposable and reusable PPE we contend that health care systems should consider moving toward a PPE strategy weighted more toward maximum per worker protection with reusable alternatives.

8 Strategy for PPE Procurement to Account for Pandemic

We assert that a viable strategy must include some element of reusable PPE for each health care worker. Assume that there is a reusable piece of PPE (PAPR suit, Hazmat suit) that can fully protect against any infectious disease. Calculate the cost of procuring one of these suits for every health care worker in a given system. Each health care worker would then be insured against the worst-case scenario. No matter the size nor the timing of the next pandemic, the PPE problem for the health care workforce would be addressed.

Standard issue of PAPR systems for all healthcare system personnel is analogous to standard practice of police, military and firefighter equipment issue strategies. When police officers graduate from the police academy they receive a standard issue firearm. When soldiers graduate from boot-camp they receive standard body armor. Likewise firefighters receive a helmet, ax and boots. By law OSHA requires employers to supply proper PPE to employees free of charge [1]. However costs cannot be passed through directly to consumers as these charges are typically not billed under current reimbursement arrangements. Regulatory bodies and health systems may have to reconsider how these costs are budgeted and apportioned if they want to mitigate shortage issue with PPE in the future.

In order for a hospital system to be prepared for a highly complex event such as a pandemic it must create a simplified approach to ensuring that every employee has access to PPE. This concept massively reduces the complexity of predicting the number of pieces of equipment that a health system must procure. This can be done at the time of hire. In this scenario the hospital system is essentially paying more upfront for reusable respirators so that in the case of a disaster scenario, they will not have to pay the premium that comes when supply chains are disrupted and prices rise dramatically. Nor will they be as likely to have to reduce capacity due to workers going on sick leave. Additionally they will also insure against total loss in the sense that disposables may not be available at all.

9 Conclusion

Pandemics fall under a unique category of risk. Their effects are multiplicative and statistically fall into the fat-tailed domain. They therefore require a unique set of statistical tools to analyze and deal with them.

It is not possible to predict the size of the next pandemic. Strategies to prepare the population in general and health care workforce more specifically must take this unique property into account and adjust preparations strategies accordingly.

We advocate a shift in mindset away from attempting to predict the size of the next pandemic and thus size of the stockpiles of disposable PPE necessary to cover the health care workforce. We instead recommend reorienting toward a model, which focuses on the system's overall exposure to risk and plan accordingly with more maximum protective, reusable PPE.

This means that when possible, teams that source PPE should bias their strategy toward disposables in equilibrium states and toward reusable PPE(purchased during equilibrium states) during times of crisis. To do this effectively health systems must plan ahead since it is not a matter of if the next pandemic will arrive, but when it will do so.

Bibliography

[1] Cohen J, Rodgers Y van der M. Contributing factors to personal protective equipment shortages during the COVID-19 pandemic. Prev Med. 2020;141:106263.

[2] Brooks, Chris https://labornotes.org/2020/03/using-trash-bags-gowns-interview-new-york-nurse

[3] Rowan NJ, Laffey JG. Challenges and solutions for addressing critical shortage of supply chain for personal and protective equipment (Ppe) arising from Coronavirus disease (Covid19) pandemic - Case study from the Republic of Ireland. Sci Total Environ. 2020;725:138532.

[4] https://www.cdc.gov/coronavirus/2019-ncov/hcp/ppe-strategy/index.html

[5] Wu HL, Huang J, Zhang CJP, He Z, Ming WK. Facemask shortage and the novel coronavirus disease (COVID-19) outbreak: Reflections on public health measures. EClinicalMedicine. 2020;:100329.

[6] DiPaola C, DiPaola M. The Case for Standard, Contact and Airborne Precautions for All Health care Workers During COVID 19 Pandemic. May 2020. ResearchersOne.

[7] Kamerow, D BMJ 2020; 369 m 1367. 3 April 2020

[8] T. Burki. Global shortage of personal protective equipment Lancet Infect. Dis., 20 (7) (2020), pp. 785-786

[9] Rawls, Carmen G., and Mark a. Turnquist. 2010. "Pre-Positioning of Emergency Supplies for Disaster Response." Transportation Research Part B: Methodological 44 (4). Elsevier Ltd: 521–34. doi:10.1016/j.trb.2009.08.003.

[10] Gooding, Emily J. (Emily Joanne). A Mixed Methods Approach to Modeling Personal Protective Equipment Supply Chains for Infectious Disease Outbreak Response. Massachusetts Institute of Technology, 2016. dspace.mit.edu, https://dspace.mit.edu/handle/1721.1/104810.

[11] Hong, Xing, Miguel a. Lejeune, and Nilay Noyan. 2015. "Stochastic Network Design for Disaster Preparedness." IIE Transactions 47 (4): 329–57. doi:10.1080/0740817X.2014.919044.

[12] Cirillo, P., Taleb, N.N. Tail risk of contagious diseases. Nat. Phys. 16, 606–613 (2020)

[13] Taleb N.N., Statistical Consequences of Fat Tails. STEM Academic Press. 2020

[14] L. de Haan, A. Ferreira (2006). Extreme Value Theory: An Introduction. Springer.

[15] Taleb, Nassim Nicholas, 1960-. (2007). The black swan : the impact of the highly improbable. New York :Random House

[16] Bandyopadhyay S, Baticulon RE, Kadhum M, et al. Infection and mortality of healthcare workers worldwide from COVID-19: a systematic review. BMJ Glob Health. 2020;5(12):e003097.

[17] Ramsey, Stuart 2020 https://news.sky.com/story/coronavirus-the-italian-covid-19-hospital-where-no-medics-have-been-infected-11966344

[18] Kressel AB, McVey JL, Miller JM, Fish LL. Hospitals learn their collective power: an isolation gown success story. Am J Infect Control. 2011;39(1):76-78.

[19] Baker N, Bromley-Dulfano R, Chan J, et al. Covid-19 solutions are climate solutions: lessons from reusable gowns. Front Public Health. 2020;8:590275.

Policy Responses to the Nonlinear Future of COVID-19's Aftermath

Anant Jani[1,*], Yohei Kawazura[2]

[1]Oxford Martin School, University of Oxford, UK.
[2]Frontier Research Institute for Interdisciplinary Sciences, Tohoku University, Japan.
*Corresponding author: anant.r.jani@gmail.com

COVID-19's nonlinear aftermath will be impossible to predict or to reverse. Despite this, we must act to minimize the damage, and potential for damage, to our citizens and society. When deciding on the best actions to take, we must distinguish between situations of risk versus fundamental uncertainty – for risk-based decisions we have knowledge about how different variables interact and can accurately and robustly measure the impact of our interventions whereas in situations of fundamental uncertainty, the outcomes of our actions are unpredictable and statistical analyses cannot produce reliable probability estimates.

Many decisions post-COVID-19 will be in situations of fundamental uncertainty; in these types of situations it will be best to follow simple rules known as heuristics and our focus should be on simple actions on key areas that could cause our citizens and societies to suffer if left unaddressed – namely, social determinants of health, which account for 80 percent of health outcomes.

1 Introduction

COVID-19 has negatively affected the livelihoods of millions of people across the world. We are seeing unprecedented impacts on key areas of our society such as health, employment and education despite the best efforts of actors across all sectors to mitigate the damage caused by the pandemic (Box 1).

Box 1. Acute negative impacts of the COVID-19 pandemic

Economy & Employment

- There was a 5.2% contraction in global GDP in 2020, which is the largest global recession in decades and the fastest and most severe downgrade in growth projections since 1990 [1]

- Economies in the Global North are expected to shrink 7-8% [1].
- Millions of jobs have been lost around the world and the global unemployment rate could increase from 4·936%-5·644% [2,3].
- Per capita income has also contracted globally and in the largest proportion of countries since 1870 [1].
- Up to 300 million people internationally could have fallen below the poverty line and 70-100 million people could have fallen into extreme poverty in 2020, which would undo progress made since 2017. This would represent an increase of 2.3% in the poverty rate compared to a no-COVID-19 scenario [4,5].
- Global debt has increased by $24 trillion over the last year, with no signs of near-term stabilisation. This level of debt is much higher than what was seen during the 2008 global financial crisis [6,7].

Food & Nutrition

- Global food prices rose by ~20% (January 2020-January 2021), which combined with reduced incomes means that households will have to decrease the quantity and quality of food they are consuming [8].
- Country surveys across dozens of countries indicated a significant number of households (up to 40%) are running out of food or reducing their consumption with someone skipping at least one meal in an average of half of households in the poorest countries [4,8].
- The total number of acutely food insecure people across 79 countries was expected to increase to 272 million by the end of 2020 [8].
- effects of these disruptions on maternal and under-5 child deaths in 118 low-income and middle-income countries; reductions in coverage and use of maternal and child health services could lead to an excess 1157000 child deaths and 56700 maternal deaths over 6 months [2].

Education

- The pandemic represents the largest disruption to education systems in history and has affected over 1.6 billion learners in over 190 countries across all continents, representing 94% of the world's student population and up to 99% in low and middle income countries [9–12].
- Lack of access to internet and digital technologies means that many children, especially those of poorer households, will fall further behind and the proportion of children below minimum education proficiency will likely increase by 25% [10,13].
- School closures also affects the provision of other essential services and benefits to families (e.g. access to nutritious food, ability of parents to work, increased violence against women and girls) [11,12].

Mental health & Substance abuse

- Individuals with existing mental illness have experienced a detrimental impact on their mental health and in some countries we are seeing a two fold increase in the number of adults experiencing some form of depression [14,15].

- Increases unemployment could be associated with an increase in suicides of about 9570 per year [3].

- The combination of unemployment, increased financial difficulties, social isolation, uncertainty about the future, and disruption to clinical services could contribute to increased alcohol intake [16–18].

- The pandemic has also negatively affected mental health services in 93% of 130 countries surveyed [19].

Evidence from previous pandemics and crises suggest that COVID-19's aftermath will be devastating and will take us years to recover from with every aspect of our societal and global systems being affected (Box 2). Because of the nonlinear nature of these changes, we know that we will not be able to predict the trajectory of the aftermath or reverse our systems to their pre-COVID states [20,21].

Box 2. Expected medium-long term negative impacts of the COVID-19 pandemic

Economy & Employment

- The deep recessions triggered by the pandemic are expected to have medium-long term effects because of lower investment, fragmented global trade and decreases in human capital because of lost work and education [1].

- The uncertainty regarding the trajectory of COVID-19 will affect the global economic outlook, which means that high unemployment rates will likely recover slowly. This was also seen with the 2008 global financial crisis where unemployment rates took seven years to return to pre-2008 levels [22].

- Job losses generally have long-lasting effects on the employment, earnings, and income prospects of laid-off workers and can negatively impact communities as well [23].

- Youth unemployment is particularly problematic and high levels are expected post-COVID, which was a trend also seen during the 2008 global financial crisis. Several years passed before youth unemployment rates matched or went below pre-crisis levels [24].

- Youth unemployment has irreversible consequences linked to the 'scarring effect', including permanently lower earnings by ~1.2% per

---

year for each additional month of unemployment and an increased probability of being unemployed later in life [24,25].

Food & Nutrition

- Simulations suggest ~0.6 years of schooling globally will be lost due to school closures, with a large proportion occurring for children of lower socio-economic status [4].
- ~24 million additional children and youth may drop out or not have access to school in 2021 because of the pandemic's economic impact [11–13].
- Many children are at risk of never returning to school, which would undo years of progress [10].
- Learning from past disaster that disrupted schooling, we know that effects on education can extend beyond this generation and produce differences observable years later [4,11–13].
- Models suggest that students currently in school may lose $10 trillion in earnings over their work life if schools are closed for five months [13].

Education

- Learning from previous crises, we know that the Indirect effects of COVID-19 will have long-term health consequences for individuals and society [2].
- Unemployment and job insecurity is linked with several negative health outcomes including all-cause mortality, death from cardiovascular disease and suicide, and higher rates of mental distress, substance abuse, depression, and anxiety [9,24–26].
- Nutritional deprivation of children and mothers can have long-term negative consequences including detrimental impacts for cognitive development of young children [4,8].

Health

- Learning from previous crises, we know that the Indirect effects of COVID-19 will have long-term health consequences for individuals and society [2].
- Unemployment and job insecurity is linked with several negative health outcomes including all-cause mortality, death from cardiovascular disease and suicide, and higher rates of mental distress, substance abuse, depression, and anxiety [9,24–26].
- Nutritional deprivation of children and mothers can have long-term negative consequences including detrimental impacts for cognitive development of young children [4,8].

Inequalities

- Evidence from previous pandemics suggest that we will see greater increases in inequality and reduced social mobility with larger welfare impacts and slower recovery for the poorest households [4].

- Income inequality increased over five years following previous pandemics from 2003-16 with the effects being higher when there were also negative effects to the economy, as we are seeing with COVID-19 [4].

- Negative physical and mental health impacts of unemployment will likely be felt more for those of lower socio-economic status and those with lower skill-sets, as seen with the 2008 global financial crisis [25].

- The full impacts of inequality go beyond five-year post-pandemic because inter-generational effects [4].

In light of these daunting projections and uncertainty about the future, what should policy makers and society do?

2 The world as a complex system

The world can be seen through the lens of systems, with a 'system' defined as:

> ...a set of things – people, cells, molecules, or whatever – interconnected in such a way that they produce their own pattern of behaviour over time. The system may be buffeted, constricted, triggered, or driven by outside forces. But the system's response to these forces is characteristic of itself, and that response is seldom simple in the real world [27].

This definition reveals one of the key requirements of viewing the world as a system – namely, accounting for its complexity and as something:

> ...made up of many elements which, as a consequence of mutual cooperation, exhibit a phenomenology that is difficult to predict. The elements and rules by which they interact may be considered well known, however, it is far from easy to explain the emergent properties at a higher level of observation as a consequence of the properties of the elements at a lower one [20].

Simplified, this can be understood as the classic saying of "the sum is greater than the parts", which means we cannot expect simple, linear relationships between the different elements of a complex system. Instead, the elements behave in a non-linear fashion with complex system non-linearity characterised as being:

> ...subject to irreversibility, such that given some change in the inputs to the system, undoing the change does not necessarily return the system to its start, whereas all linear systems are reversible. Furthermore nonlinear systems can be subject to discontinuous or catastrophic state changes[...] Managing such systems, particularly in response to some sort of failure is very difficult [21].

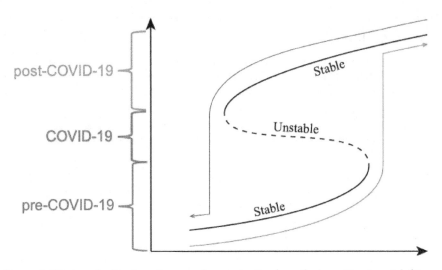

Figure 1: **Hysteresis.** Hysteresis curve demonstrating how the transitioning stability from pre-post COVID-19 will lead to a new stable state. The curve is a fixed point of a nonlinear equation with the solid indicating a stable state and the dashed line corresponding to an unstable state.

A classic example of non-linearity within complex systems is traffic jams. Car speed on a highway is only slightly affected over a large range of car density but the density will have a tipping point beyond which even a small increase in car density will lead to a disproportionate decrease in traffic flow, leading to a traffic jam [27].

COVID-19 in the context of a complex system

Acknowledging that complex systems may not predictable or reversible, it is important to understand the implications of this in the context of systemic crises like the COVID-19 pandemic and its aftermath [27]. The reversibility of a complex system is linked to the concept of hysteresis, which is dependence of a state on its history [28]. If we imagine a path of a state in parameter space, a stable pre-COVID-19 state (Lyapunov exponent <= 0) gradually approaches the unstable COVID-19 state (Lyapunov exponent > 0) and at some point in the future, the state will be stabilized to the post-COVID-19 state. Once the state has reached the post-COVID-19 state, it will not be possible to go back to the pre-pandemic state easily by changing the parameters – all societies will need to face a new 'normal' (Fig 1). It is also important to note that a stable state does not mean that it is fair, equitable or equally beneficial for all elements of the system – a stable state only indicates that it is not subject to the unpredictable fluctuations one sees in response to a systemic shock like the slight increase in car density that leads to a traffic jam or a global systemic shock like the COVID-19 pandemic.

3 Decision making in situations of risk vs. fundamental uncertainty

The unpredictability of our post-COVID-19 future is unsettling but it also provides us with a sense of hope because we can strive to build a more fair, equitable, resilient and prosperous world post-COVID-19. Building our post-COVID-19 future requires that critical decisions be made about various aspects of our society and for each decision, we must determine whether the decisions will be made in situations of risk or fundamental uncertainty. Decision-making in situations of risk or fundamental uncertainty require different approaches and are context-dependent. In situations of risk, we have knowledge about how different variables interact and also have the ability to accurately and robustly measure the impact of our interventions. In situations of fundamental uncertainty, the outcomes of our actions are unpredictable and too unique to allow for statistical analyses that can yield reliable probability estimates. Two important things to note about fundamental uncertainty are that it is context-dependent and it is not static. The context is important because though knowledge about how different variables interact and methods to collect data on these interactions exists, there will be heterogeneity in the distribution of this knowledge as well as the capacity to collect data. Furthermore, fundamental uncertainty is not static because it can be reduced to risk with research, knowledge and novel methods of collecting data, which gives decision-makers the opportunity to make risk-based decisions [29].

Situations of risk or fundamental uncertainty require different decision-making methods. A tendency in these days of big data, machine learning and artificial intelligence is to harbour an expectation that the more data we have the better. The use of greater amounts of data are called for in situations of risk where we know the consequences of the options available to us but in situations of fundamental uncertainty where the outcomes of our actions are unpredictable, calculations with large amounts of data and complex algorithms gives decision-makers a false sense of security and normally have limited benefit in helping decision makers make better decisions because of the phenomenon of overfitting, which was aptly demonstrated by Google Flu Trends [29, 30]. In situations where decision-makers are faced with fundamental uncertainty, simple approaches, known as heuristics, are the appropriate option. Heuristics are strategies for decision-making adapted to the decision-maker's local context, which can reduce effort and lead to more accurate judgements by ignoring complexity and avoiding overfitting. Some classic examples of heuristics are satisficing the 1/N rule [29].

4 Heuristics for COVID-19's aftermath

At a system's level, COVID-19's aftermath will be nonlinear and, therefore, fundamentally uncertain. Despite this, we will need to act to reduce the negative impacts on our citizens and, ideally, build a more sustainable and resilient post-COVID-19 world. As a starting point, a heuristic approach we can adopt can focus on the key sources of suffering for our citizens in COVID-19's aftermath as well interventions that can be implemented to reduce the suffering of our citizens in the future while also building up the resilience of our society to be less negatively affected by future shocks we will face.

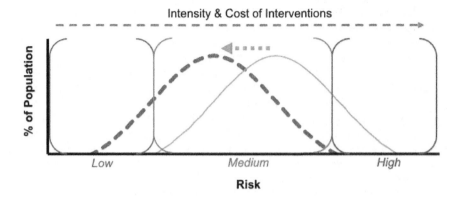

Figure 2: Shifting risk profiles to promote population health. Rose's model of improving population health by shifting risk profiles to the left (e.g. from the yellow to blue dashed bell curve). Overlaid into the different risk strata are the level of interventions needed to keep the strata healthy (red arrow above). Figure adapted from references [31, 32].

The heuristic governments are currently taking is to pump money into their economies to help businesses and citizens, which is the correct approach but it is important to note that the trillions being injected into global systems are necessary but not sufficient now or in the future to fully address the suffering our citizens will face [6, 7].

We have known for many decades that social determinants health (e.g. housing, education, transportation, access to jobs, nutritious food, clean air, clean water, support to prevent/recover from substance abuse, etc.) determine 70-80% of health outcomes [31–33]. Supporting mechanisms to address these factors individually and collectively will go a long way to prevent or reduce suffering of our citizens as well as to reduce the level of resource needed to address their needs now and in the future. To promote population health, our systems need to move away from focusing on high risk individuals to shifting the entire risk profile of our populations to the left by addressing the factors that have the greatest influence on the incidence of morbidity and mortality – e.g. social determinants of health (Fig 2) [31, 32].

Shifting the population risk profile to the left is even more important in light of COVID-19's aftermath because many of our citizens will transition to states of poor health [2, 4, 8, 23–26]. It will be almost impossible to predict the 'phase transitions' our individuals citizens make from health to disease when they move low to medium/high risk or medium to high risk [28, 34]. Furthermore, once in the relatively higher risk phase, the level of intervention needed for those individuals will be much higher than if we intervened at a population level to control the incidence of their particular morbidity.

5 A heuristically-informed quincunx for COVID-19's aftermath

With so much money being allocated into our systems and the acknowledgement of the need for change, now is the time when we can introduce simple heuristic approaches to rearrange, augment and bolster structural elements of our societies to promote the health of our populations and also increase our resilience. As Taylor-Robinson and Kee point out, we can view the elements creating our societal structures through the lens of a quincunx, which is a device where marbles funnelled into the top of the device hit an array of pins and demonstrate how a normal distribution can be generated from a random process [35]. With this perspective in mind, we can actively push to invest in simple heuristic-led strategies to address social determinants of health by, for example, ensuring our citizens have access to nutritious food (e.g. subsidising healthy foods and supporting community gardens), clean water (e.g. through state-led interventions to create or augment water purification plants/distribution networks and investing in or subsidising interventions to improve water efficiency in agriculture), jobs training (e.g. through subsidies for vocational and technical skills training), employment (e.g. the state taking the role, in the short-medium term, of an employer of last resort [36] and supporting infrastructure, preferably green infrastructure, projects).

Box 3 highlights some specific policy suggestions for three specific areas (Economy and Employment, Education and Health), for which we have evidence of positive outcomes that could be generated for citizens and society based on evidence from previous crises.

Box 3. Toolbox of policy responses to create a better and more equitable post-COVID steady state

We know from previous crises that there is a great risk for increased inequalities post-crisis. Policies should be designed to support sustainable and inclusive recovery that addresses current needs and reduces vulnerability to future crises [4]. Previous crises have taught us that maintaining and increasing public spending and social safety nets can have dramatic effects on recovery trajectories, health and dimensions of inequality [4,25,26].

Economy & Employment

Countries should focus on supporting economic activity and the drivers of economic growth while also providing support for households, businesses and essential services [1].

- The negative long-term social and health-related effects of unemployment could be mitigated through initiatives like universal basic income, temporary income support for displaced workers (e.g. unemployment insurance, redundancy payments, social assistance programs), implementing national job guarantee programs (e.g. with governments functioning as an Employer of Last Resort and supporting public works programs such as green infrastructure projects) and targeted active labour market programs (e.g. employment services such as labour exchanges, education and training, business support or subsidized employment programs as well as access to affordable childcare and improved parental leave policies) [4,23,26,36].

- Improved access to low-cost financial products and improving digital infrastructure could also help micro and small enterprises survive and become more competitive [4].

Education

Countries should generally focus on three areas for educational recovery - coping, managing continuity and improving/accelerating improvements to the educational system to improve outcomes, addressing inequality, and reducing learning poverty [13].

- Coping and Managing Continuity: in the short-medium term, school capacity should be strengthened to reduce risks of disease transmission and promote healthy behavior and re-enrolment campaigns and cash transfers should be instituted to ensure schools do not lose children permanently to drop-out, particularly vulnerable group and students below learning proficiency standards [4,13].
- Improving/accelerating improvements: Investments should be made to support teacher training as well as in bolstering school infrastructure through technology-enhanced learning [13].

Health

Countries should ensure there are sufficient safety nets and social welfare programs to ensure individual and community health needs can be met.

- Core infrastructure components including social registries and mobile payment systems to identify vulnerable individuals to ensure they do not slip through the cracks as well as tracking emerging risks should be implemented [4].
- Programs to reduce food insecurity (e.g. school meals, food subsidies, etc.) will be essential to ensure we avoid short, medium and long term negative consequences of malnutrition
- Increased mental illness can be expected post-pandemic and steps should be put in place early to prepare for this by raising awareness and bolstering services such as hotlines and psychiatric services [3].

These three areas are highlighted as a starting point because addressing them has the added benefit of tackling other social determinants – for example, strengthening employment opportunities for individuals so that they are able to earn a livelihood means that they are more likely to be able to afford nutritious food, adequate housing and transport and the employment status of the individual will protect them from adverse mental health outcomes.

COVID-19's aftermath will be unpredictable but a bleak future post-COVID is not a foregone conclusion. It will take time, new ways of thinking and new ways of working but it is fully in our power to design and build a better future if we make the right types of decisions and focus on the elements, like social determinants of health, that can build a stronger present and future for our citizens.

Bibliography

[1] World Bank. The Global Economic Outlook During the COVID-19 Pandemic: A Changed World. See: https://www.worldbank.org/en/news/feature/2020/06/08/the-global-economic-outlook-during-the-covid-19-pandemic-a-changed-world (last checked 18 Feb 2021).

[2] The Lancet Global Health Editors. Food insecurity will be the sting in the tail of COVID-19. The Lancet Global Health 2020; 8: E737.

[3] Kawolh W and Nordt C. COVID-19, unemployment, and suicide. The Lancet Psychiatry 2020; 7: P389-390.

[4] World Bank Blogs. What COVID-19 can mean for long-term inequality in developing countries. See: https://blogs.worldbank.org/voices/what-covid-19-can-mean-long-term-inequality-developing-countries (last checked 18 Feb 2021).

[5] The World Bank. Projected poverty impacts of COVID-19 (coronavirus). See: https://www.worldbank.org/en/topic/poverty/brief/projected-poverty-impacts-of-COVID-19 (last checked 18 Feb 2021).

[6] Institute of International Finance. Global Debt Monitor. See: https://www.iif.com/Research/Capital-Flows-and-Debt/Global-Debt-Monitor (last checked 18 Feb 2021).

[7] Reuters. COVID response drives $ 24 trillion surge in global debt: IIF. See: https://www.reuters.com/article/idUSKBN2AH285 (last checked 18 Feb 2021).

[8] The World Bank. Food Security and COVID-19. See: https://www.worldbank.org/en/topic/agriculture/brief/food-security-and-covid-19 (last checked 18 Feb 2021).

[9] The World Bank. World Bank Education COVID-19 School Closures Map. See: https://www.worldbank.org/en/data/interactive/2020/03/24/world-bank-education-and-covid-19 (last checked 18 Feb 2021).

[10] UNICEF. Education and COVID-19. See: https://data.unicef.org/topic/education/covid-19/ (last checked 18 Feb 2021).

[11] The UN. Policy Brief:Educationduring COVID-19and beyond. See: https://www.un.org/development/desa/dspd/wp-content/uploads/sites/22/2020/08/sg_policy_brief_covid-19_and_education_august_2020.pdf (last checked 18 Feb 2021).

[12] UNESCO. Adverse consequences of school closures. See: https://en.unesco.org/covid19/educationresponse/consequences (last checked 18 Feb 2021).

[13] World Bank Blogs. Learning losses due to COVID19 could add up to $ 10 trillion. See: https://blogs.worldbank.org/education/learning-losses-due-covid19-could-add-10-trillion (last checked 18 Feb 2021).

[14] Pan K-Y, Kok AAL, Eikelenboom M, Horsfall M, Jorg F, Luteijn RA, et al. The mental health impact of the COVID-19 pandemic on people with and without depressive, anxiety, or obsessive-compulsive disorders: a longitudinal study of three Dutch case-control cohorts. The Lancet Psychiatry 2020; 8: P121-129.

[15] Office for National Statistics. Coronavirus and depression in adults, Great Britain: June 2020. See: https://www.ons.gov.uk/peoplepopulationandcommunity/wellbeing/articles/coronavirusanddepressioninadultsgreatbritain/june2020 (last checked 18 Feb 2021).

[16] Kim JU, Majid A, Judge R, Crook P, Nathwani R, Selvapatt N, et al. Effect of COVID-19 lockdown on alcohol consumption in patients with pre-existing alcohol use disorder. The Lancet Gastroenterology & Hepatology 2020; 5: P886-887.

[17] Finlay I and Gilmore I. Covid-19 and alcohol—a dangerous cocktail. BMJ 2020; 369: m1987.

[18] Sun Y, Li Y, Bao Y, Meng S, Sun Y, Schumann G, et al. Brief Report: Increased Addictive Internet and Substance Use Behavior During the COVID-19 Pandemic in China. Am J Addict 2020; 29: 268-270.

[19] WHO. COVID-19 disrupting mental health services in most countries, WHO survey. See: https://www.who.int/news/item/05-10-2020-covid-19-disrupting-mental-health-services-in-most-countries-who-survey (last checked 18 Feb 2021).

[20] Marro J. Physics, Nature and Society: A Guide to Order and Complexity in our World. Switzerland: Springer, 2014.

[21] Fieguth P. An Introduction to Complex Systems: Society, Ecology and Nonlinear Dynamics. Cham, Switzerland: Springer, 2017.

[22] Bank of England. How persistent will the impact of Covid-19 on unemployment be?. See: https://www.bankofengland.co.uk/bank-overground/2020/how-persistent-will-the–impact-of-covid-19-on-unemployment-be (last checked 18 Feb 2021).

[23] World Bank Group. Causes and Impacts of Job Displacements and Public Policy Responses. See: https://openknowledge.worldbank.org/bitstream/handle/10986/33720/Causes-and-Impacts-of-Job-Displacements-and-Public-Policy-Responses.pdf?sequence=1& isAllowed=y (last checked 18 Feb 2021).

[24] Grzegorczyk, M. and G. Wolff (2020) 'The scarring effect of COVID-19: youth unemployment in Europe', Bruegel Blog, 28 November

[25] Bambra C, Riordan R, Ford J and Matthew F. The COVID-19 pandemic and health inequalities. J Epidemiol Community Health 2020; 74: 964-968.

[26] Hensher M. Covid-19, unemployment, and health: time for deeper solutions? BMJ 2020; 371: m3687.

[27] Meadows D. Thinking in Systems. Vermont, USA: Chelsea Green Publishing, 2008.

[28] Thom, René. Structural Stability and Morphogenesis: An Outline of a General Theory of Models. Reading, MA: Addison-Wesley, 1989.

[29] Mousavi S and Gigerenzer G. Risk, uncertainty, and heuristics. J Bus Res 2014; 67: 1671–1678.

[30] Lazer D, Kennedy R, King G and Vespignani A. The parable of Google Flu: Traps in big data analysis. Science 2014: 343: 1203-1205.

[31] Rose G. Sick individuals and sick populations. Int J Epidemiol 1985; 14: 32–38.

[32] Rose, G. A., Khaw, K.-T., & Marmot, M. G. (2008). Rose's strategy of preventive medicine: The complete original text. Oxford, UK: Oxford University Press.

[33] WHO. Social Determinants of Health: the Solid Facts, 2nd edn. Geneva: WHO, 2003.

[34] Kawazura, Y., & Yoshida, Z. Entropy production rate in a flux-driven self-organizing system. Physical Review E 2010; 82: 066403.

[35] Taylor-Robinson D and Kee F. Precision public health – the Emperor's new clothes. International Jounral of Epidemillogy 2019; 48: 1-6.

[36] Minsky HP. Ending Poverty: Jobs, not welfare. New York, NY: Levy Economics Institute, 2013.

Collective Intelligence and Governance for Pandemics

David Pastor-Escuredo[1,2,*], Carlota Tarazona-Lizarraga[2], Annalyn Bachmann[3] and Philip Treleaven[1]

[1]UCL, London.
[2]LifeD Lab, Madrid, Spain.
[3]MIT Centre for Collective Intelligence, MIT, Cambridge, USA.
*Corresponding author: david@lifedlab.org

Pandemics are a complex problem that involve all societal mechanisms and is related to heterogeneous human behaviors and also collective patterns. Remarkably, even when local systems may greatly differ from each other, pandemics propagate because of the pathways that connect the different systems and several invariant behaviors and patterns that have emerged globally. These structures and properties make the current world fragile against pandemics. A necessary and proper reconfiguration of prevention and response systems for pandemics should be addressed based on complexity, ethic and multi-scale systems so that the world becomes more robust, resilient and anti-fragile without losing the advantages of being interconnected. A virus is a rather simple biological unit that, given the probabilistic combinations of biochemical mechanisms and human behaviors, is capable to propagate using humans as a substrate. Viruses exploit complexity and emergence to turn into epidemics. A top-down approach led by Governmental organs for managing such a phenomenon is not sufficient and may be only effective if policies are very restrictive and their efficacy depends not only in the measures implemented but also on the dynamics of the policies and the population perception and compliance. This top-down approach is even weaker if there is not a national and international coordination capable of fighting back the scalability and massive and fast propagation of pandemics. Combining coordinated top-down measures with the right timing must be complemented with bottom-up approaches to generate a collective response. A collective response includes behavioral changes regarding hygiene and physical distancing, but also the transmission of trustful information and the generation collective local efforts and constructive perception and sentiment. Such a collective response would reinforce policies to be more impactful and have faster and more effective dynamics. In an age of passive synchronization driven by digitalization, active collective action and response is needed for building up robustness, resilience, anti-fragility and ethical response and recovery. Collectiveness can hardly emerge without signaling, sensing and leadership mechanisms. Here, we make a commentary about potential of multi-scale collective response to pandemics and present a framework to understand social organization in information transmission based on social networks analysis during COVID-19 for discussion and recommendations.

1 Introduction

Pandemics are not just a biological phenomenon, but also a socio-economic and cultural one, both in terms of their nature and their impact. Consequently, digital epidemiology has appeared as a discipline itself that powered by new data sources such as mobile phone data and social network is capable to deciphering and quantifying the dynamics of the propagation of epidemics [1]. Network science has been applied for characterizing and modelling disease spreading considering different types of network topology and dynamics [2–7]. Furthermore, the availability of geo-localized data has enabled assessing the contribution of mobility to epidemic spreading and socio-economic impact with fine detail [8,9].

The simplicity of viruses, that through combinatorial mutations may find a way to reproduce massively, use not only the human body to survive and reproduce, but the entire society as a substrate for transmission. The tangled network created through digital connectivity, world-wide economic and financial exchanges and multi-scale travelling patterns results in a very transmissible substrate for new viruses to turn into pandemics. Each epidemic has its own characteristics in terms of immunology and virology. For instance, malaria affect mainly children that higher risk of death under malnutrition conditions [9]. Influenza is another disease where children play a key role in transmission [10–12]. On the contrary, children have shown better immunology response to COVID-19 although the transmission patterns and infectious power vs viral load profiles are yet to be properly characterized [13–15]. The characterization of transmission patterns is complex and requires a local and systematic analysis based on clinical studies, surveys and Big Data to create contact matrices that help building predictive models and monitoring systems [16–24].

However, regardless the specificities of the viruses and the local transmission chain, pandemics share some commonalities that turn them into high-speed and scalable propagation diseases and they seem to have found their way through our world-wide structure. Additionally, there exists hypothesis that the global industrial system is exerting an unbearable load of stress to the planet leading to less self-regulation and richness of biodiversity and, therefore, increasing our exposure to viruses of different kind and the risk of more frequent pandemics. The social dimension and the biological dimension of the planet are deeply interconnected and epidemics are resulting phenomena that may be more frequent and impactful in the years to come.

This situation poses the reasonable question whether the globalized world is itself, by definition, weak and fragile and if pandemics will change the world and threat humankind and civilization. It is important to highlight that while epidemics do not intrinsically account for socio-economic profiles of individuals for transmission, there are population groups that may have higher exposure due to problems to adapt to lockdowns, mobility restriction and other distancing and non-pharmaceutical measures [18, 22]. The hygiene, water and sanitation conditions

(WASH) have a very relevant role in epidemics, both on their onset and their tail so we can expect poor WASH regions to become reservoirs of diseases such as COVID-19 in the same way they are for another viruses. Demographics and urban-rural development will also play a key role on long-term transmission chains. Furthermore, most part of the enduring socio-economic impact of pandemics will affect the most vulnerable population because of the deterioration of the local and global economy and regional livelihoods and the changes in the commerce and consumption patterns.

Thus, pandemics are a multi-scale processes that affect all layers of the social tissue at local and global levels and have a modulated impact depending on the vulnerability of the people. This ubiquity and mixed homogeneity and heterogeneity of pandemics result in a problem that is very hard to model and predict. Pandemics are socio-biological anomalies that can be considered fat-tailed [25, 26] and cannot be controlled with existing governance mechanisms. COVID-19 has led to an intensive digitalization for communication, work and also epidemics management with a surge of uncountable models, tools and systems [27]. This digitalization may have mixed consequences as there is not an underlying scientific framework to use these tools, neither ethical and governance ones. Existing predictive mechanisms have also limited rigor because of the lack of all the necessary data in real-time (apart from embodied assumptions into the models that would need scientific contrast) and therefore should be used under very well-thought conditions as the negative outcomes of wrong data-drive decision making for pandemics can have devastating impact [28].

We can foresee the negative impact of digital platforms and wrong evidence-based system from the information spreading patterns in the networks. Misinformation and fake news are a recurrent problem of our digital era [29–31]. The volume of misinformation and its impact grows during large events, crises and hazards [32]. When misinformation turns into a systemic pattern it becomes an infodemic [33,34]. Infodemics can amplify the real negative consequences of the pandemic in different dimensions: social, economic and even sanitary. For instance, infodemics can lead to hatred between population groups [35] that fragment the society influencing its response or result in negative habits that help the pandemic propagate.

Harnessing digital systems for pandemics prediction and response need a framework of complexity integrating governance, multi-scale drivers and collective intelligence. Here we discuss several aspects that should be considered for such a framework and provide several recommendations based on learning from social science and also developmental biology and biophysics that inspire the design of resilient systems. In this light, we propose the integration of technology and data and AI-driven systems. We also present a social network analysis to provide a discussion of real network structures that can affect the response to COVID-19 and the pace of the recovery and adaptation to potential endemic pandemics.

2 Collective Intelligence and Action

People play an unaware role in pandemics, being the source of incubation, reproduction and transmission of viruses. Non-pharmaceutical measures point in several case to individual-level actions being using mask, staying home or washing hands. Small preventive hygiene actions have a direct impact in the epidemiological evolution as they have a scalable inhibiting effect against the virus propagation. High infectious diseases such as COVID-19 still propagate even when hygiene measures are taken and enforced by most part of the population. When the epidemiological curves passes certain thresholds, more restrictive measures may be enforced by Governments limiting people's freedom and having a severe socio-economic impact.

These measures are taken towards cutting the individual level pathways for the transmission reducing the average reproductive number R to avoid scalable contagion network dynamics [2, 36]. However, the distribution of R (characterized by the parameter k) may be very wide leading to the appearance of superspreading events and superspreaders that hamper holding the epidemiological curve flat, allowing a progressive decay of the curve and the eventual eradication in a hypothetical isolated context. Additionally, imported cases also hamper local eradication or keeping epidemiological curves flat low.

Full lockdowns may be the only sure way to inhibit transmission pathways among individuals provided a sufficient time window to disconnect households. However, lockdowns have not be as effective along time everywhere and Governments have not been capable of sustain them long enough globally. In a context of persisting pandemic which may turn endemic as recently claimed my WHO , population will need to keep playing as individual inhibitors by hygiene and physical distancing measures. However, this may not be sufficient to prevent from secondary waves which will cause more deaths and socio-economic crises when travel restrictions disappear or people relax in physical distancing.

Population must gain awareness and have tools to act so that they can prevent themselves and their environment to suffer from the pandemic. There are little tools for people to do so and they have not been properly designed in most cases. So far, digital tracing apps have acted mainly as a control mechanism, but not as an empowering mechanism. Trust has not been at the core of digital tracing apps and they have failed in many places. Digital tools have to improve "sensing and communication" more than increase surveillance. For instance, timely and accurate information of the epidemiological curve in our residential and work environments would allow people to increase their awareness and precautions which cannot be kept at the same level all the time throughout a long lasting pandemic. In that sense, bidirectional communication with health authorities would greatly increase trust instead of generating the reasonable perception of being observed and controlled that digital contact tracing apps have generated. Besides, there are key technology design elements that have to be aligned with a (collective) human-centered design such as privacy-safety or decentralized analysis [37]. Through sensing and communication, collective response and precaution patterns can emerge without a unidirectional control system. Pandemics are a great opportunity to test how social systems can self-organize (provided the right mechanisms and monitoring) as biological tissues or animal herds do.

Furthermore, most part of tracing apps do not account for all the population [38, 39] which has ethical and epidemiological negative consequences. Reaching vulnerable population so they are properly assisted minimizing their exposure and risk would have great benefits in the epidemiological curve and the mortality. As mentioned, vulnerable population like elderly or low-income workers may have problems to be compliant with physical distancing and take precautions leading to an unfair risk of contributing to the propagation. It is necessary to design technology and guarantees that account for the asymmetries in the population in terms of information, precaution capacities and individual resilience. Having population groups that are fragile and not resilient make the whole system fragile and not resilient. By generating more trustful links among the population along the different social and demographic layers and with the decision makers and health system, we could expect to have a more robust and resilient social tissue that is able to sustain compliance through time and react faster to surges in the epidemiological curve. Loneliness is not the proper mental and epidemiological state to fight a pandemic.

Collective Intelligence and Artificial Intelligence can be even more relevant tackling the socio-economic crises derived from the pandemic. It is important to understand how the economy should be stimulated in order to combine three main drivers: accelerators, synergic actions and inclusive mechanisms. This requires a whole different approach to AI and evidence-based systems and the structure of decision making mechanisms and roles. Artificial Intelligence and Collective Intelligence should be oriented to design better Augmented Collective Intelligence [40–42] that harness the collaborative processes for sense making and deeper understanding of the complexity involved in the processes derived from the pandemic and how to find the right solutions.

Besides the cognitive and epistemological challenges that the COVID-19 derived crisis will pose into our societies and planet, there will be an important number of crossed interests from different sectors and ideological confrontation. It is rather difficult to imagine that the world can recover with the necessary pace without a collective effort and a scientific framework that allows identifying accelerators and synergies and ensuring that the new reality does not increase the socio-economic gaps and segregation in different parts of the world [43, 44]. Although there exist frameworks of citizen science and collective intelligence, it is necessary to go a step forward regarding the scientific grounds and the use of Artificial Intelligence and Data to generate ideas collectively and with a broad perspective, synthesize knowledge and ideas, make simulations for action, perform evaluations of impact and generate the warnings triggering actions. The appearance of technologies such as Blockchain are an opportunity to progress towards these frameworks of collective consensus, sense making and action.

Collectiveness is not contradictory with organization, however, new multi-scale organization systems are required to manage the complexity and state of emergency of the world we live in. Leadership and networks are fundamental to build a more responsive and anti-fragile socio-economic tissue. Leadership is required to drive change, but we need to consider two fundamental aspects in leadership: the content and purpose of ideas, projects, initiatives and narratives and also the topology of the network linked with the leading nodes. We briefly discuss these aspects in a case study of information spreading in Twitter data [45]. It is necessary to investigate further in how leadership should be configured and distributed to promote proactive and fast reactive actions and transformations.

This type of framework is not only required for pandemic management but is also necessary for innovation and drive constructive efforts in the private sector. A new type of socio-economic tissue based on collaboration that aligns collective efforts in specific challenges and missions with a proper scientific and quantitative frameworks [46] can help leading to a faster recover and develop the structure for building up resilience and anti-fragility. As mentioned, from biological systems we can learn that a certain level of specialization combined with interactions and sensing mechanisms are the foundation for emergent properties.

3   Governance

Transformative processes require drivers and catalyzers to generate phase transitions and enable structural changes. Besides the lack of catalyzers, transformations can be hindered by inhibitors that can take different shapes such as heavy bureaucracy, political interests, resistive narratives, hatred argumentation, social polarization, etc. Furthermore, most current organizations and governance structures are stiff and designed from the lens of steadiness, although public policies lack the vision of long-term objectives and missions. Current world challenged by biological, economic and environmental threats has to be progressively redesigned from the lens of dynamics and complexity. COVID-19 pandemic has also highlighted that in a hyper connected world, there is a severe lack of international coordination systems even within the European Union. Out of sync response among countries has proved useless and damaging facilitating the virus spreading and reaching parts of the world with a precarious health system causing deaths and likely leading to more poverty.

For these reasons, dynamic processes have to be led also by governments, multilateral actors and international institutions that represent democracy. Only through more dynamic governance it will be possible to build up resilience as a necessary requirement for sustainability by implementing policies, mobilizing investment, proper regulations, international agreements, etc. Resilience and anti-fragility require parts of the system to activate ahead and react to external conditions facilitating effective response and driving necessary and constructive transformations and change. We may think of such mechanisms as genes that activate and are specialized to response to specific stimuli. However, current power structures, even when legitimate, do not offer guarantees of such kind of response due to structural stiffness, overload of hierarchies and bureaucracy, uncertainty for decision making, wicked power relationships and responsibility ownership.

Digital technologies and AI must be used to improve response through better insights as discussed. Holistic frameworks powered by complex science used within multi-stakeholders ecosystems are the way forwards to create evidence-based policies and innovation governance towards the management of pandemics and the stimulation for recovery. Artificial Intelligence has been often used to get deep insights on bounded data and a transition to wide insights that open perspective should be the goal of improvements in Machine Learning, data science and visual analytics. Depth and wideness are two elements to promote in decision making processes through learning and training of decision makers and leaders. The integration of scientists on governance platforms organically, not only as expert panels, is a requirement to harness a secure future of data-driven governance. Scientists also need training on governance and the proper non-academic incentives.

Evidence-based policy making and AI-driven systems require also ethical frameworks to be acceptable and useful. For instance, COVID-19 pandemic has risen the paradigm of relying on digital contact tracing apps for individual control, contagion monitoring and also generating accurate contact matrices [47, 48]. The use of these apps brings social and ethical problems that depend on the technological design, their deployment, their governance and their application [37–39]. As discussed, human-centered design is required, but also a broader perspective of an acceptable digitalization both in terms of technology and technology governance. Typical technical values such as transparency, efficiency and trustworthiness are necessary but not sufficient to be the ground for evidence-based policy making in the digital era . Frameworks will depend, of course, on cultural particularities, but it is important to create debate and consensus of these frameworks prior an invasive digitalization.

Beyond evidence-based decision making, digital technology can help building more decentralized, responsive, flexible and accountable systems for governance. Although current debate points towards the bias of AI, probably, the most promising use of digital technology is to help building better public and private organizations to overcome human limitations [49]. Algorithmic governance can mediate in power relationships and open spaces for collective intelligence if properly designed and deployed, rather than making more obscure and less interpretable decision making systems [50]. Impact assessment of data-drive policies would be a necessary mechanism to ensure that technology does not worsen existing problems or create new ones.

Interdisciplinary, international and independent (and even decentralized) teams should help make decisions, interpret results and analyze outcomes. In that regard, COVID-19 has been a promising milestone of academic collaboration [51] that is necessary to leverage the necessary knowledge, technology and resources in a timely and effective manner. Collaboration will be necessarily layered integrating different actors as a network that needs management where digital technology should help as well. AI and data could help activate the network through signaling processes and governance automation as it has started happening in the humanitarian sector [52]. A certain level of automation in partnerships and collaboration including data sharing would relieve from responsibility burdens of decision makers to accelerate processes. This is rather controversial with some AI regulation experts, but responsibility should never lead to inaction and digitalization is an opportunity to make not only more transparent but also more actionable and committed governance, always under the supervision of ethics and accountable impact.

4   Case Study: Information Spreading Leaders

During a crisis such as the current COVID-19 pandemic, information is key as it greatly shapes people's opinion, behavior and even their psychological state [53–55]. However, the greater the impact the greater the risk [56]. It has been acknowledged from the General-Secretary of United Nations that the infodemic of misinformation is an important secondary crisis associated to the pandemic that can amplify the crisis. During a crisis, time is critical, so people need to be informed at the right time [57,58]. Furthermore, information during a crisis leads to action, so population needs to be properly informed to act right [59]. On the contrary, reliable and trustful information along with messages of hope and solidarity can be used to monitor and control the pandemic, create real-time response mechanisms, build safety nets and help promote resilience and antifragility.

To fight misinformation and hate speech, content-based filtering is the most common approach taken [33, 60–62]. The availability of Deep Learning tools makes this task easier and scalable [63–65]. Also, positioning in search engines is key to ensure that misinformation does not dominate the most relevant results of the searches. However, in social media, besides content, people's individual behavior and social network properties, dynamics and topology are other relevant factors that determine the spread of information in the different clusters and layers of the society organization skeleton [66–68]. Infodemics are frequent specially in social networks that are distributed systems of information generation and spreading. For this to happen, the content is not the only variable but the structure of the social network and the behavior of relevant people greatly contribute [33].

One of the characteristics of the current digital era is a certain level of centralization of information spreading and technological services. We are also witnessing certain level of influencers-type of movement regarding ethics of technology and data-driven policy making and governance. The transition from the COVID-19 to a potential different world more resilient and anti-fragile demands thinking about the role and nature of leadership and influencers.

Here, we present a preliminary study to characterize leaders in Twitter based on the analysis of the social graph derived from the activity in this social network [69]. Centrality metrics are used to identify relevant nodes that are further characterized in terms of users' parameters managed by Twitter [70–74]. Although this tool may be used for surveillance of individuals, we propose it as the basis for a constructive application to detect and empower users with a positive influence in the collective behavior of the network and the propagation of information [72, 75]. This is an example of how technology could be used to understand better social organization and take actions accordingly to build resilient networks.

Data

Tweets were retrieved using the real-time streaming API of Twitter using a filter of keywords. The keywords were basic terms to retrieve posts related to the pandemic 'coronavirus', 'Coronavirus', 'CoronavirusES', 'coronavirusESP', 'coronavirus', 'Coronavirus', 'covid19', 'covid19', 'Covid19', 'Covid19', 'covid-19', 'covid-19', 'COVID-19', 'COVID-19'. In total, 500.000 posts were retrieved in the time interval between April 18th and May 4th.

Method

Each tweet was analyzed to extract mentioned users, retweeted users, quoted users or replied users. For each post the corresponding nodes were added to an undirected graph as well as a corresponding edge initializing the edge property "flow". If the edge was already created, the "flow" was incremented. The network was completed by adding the property "inverse flow" (1/flow) to each edge. The resulting network featured 107544 nodes and 116855 edges.

To compute centrality metrics the network described above was filtered. First, users with a node degree (number of edges connected to the node) less than a given threshold (experimentally set to 3) were removed from the network as well as the edges connected to those nodes. The reason of this filtering was to reduce computation cost as algorithms for centrality metrics have a high computation cost and also removed poorly connected nodes as the network built comes from sparse data (retweets, mentions and quotes). However, it is desirable to minimize the amount of filtering performed to study large scale properties within the network.

Degree	Radial and volume-based centrality computed from 1-length walks (normalized degree) based on the flow property. This metric measures the number of direct connections that an individual node has to other nodes within a network.
Eigenvalue	Radial and volume-based centrality computed from infinite length walks. This metric measures the number of edges per node, and the number of edges of each connected node and so on.
Closeness	Radial and length-based centrality that considers the length of the shortest paths of all nodes to the target node based on the flow property.
Betweenness	Medial and volume-based centrality that considers the number of shortest paths passing by a target node based on the flow property. This centrality was computed for both directions of the directed graph.
Current flow Closeness (cfcloseness)	Radial and length-based centrality based on current flow model using the inverse flow property. This centrality was computed for the largest connected undirected subgraph. This metric is a variant of closeness that evaluates not only the shortest paths, all possible paths
Current flow Betweenness (cfbewteennes)	Medial and volume-based centrality based on current flow model using the inverse flow property. This centrality was computed for the largest connected undirected subgraph. This metric is a variant that evaluates the intermediary position in all the paths between the rest of nodes.
Load	The load centrality of a node is the fraction of all shortest paths that pass through that node. Load centrality is slightly different than betweenness.

Table 1: Centrality descriptors table

The resulting network featured 15845 nodes and 26837 edges. Additionally, the network was filtered to be connected which is a requirement for the computation of several of the centrality metrics described below. For this purpose, the subnetworks connected were identified, selecting the largest connected network as the target network for analysis. The resulting network featured 12006 nodes and 25316 edges.

Several centrality metrics were computed: current-flow betweenness, betweenness, closeness, current-flow closeness, eigenvalue, degree and load. Each of this centrality metric highlights a specific relevance property of a node with regards to the whole flow through the network. Descriptors explanations are summarized in Table 1. Besides the network-based metrics, Twitter user' parameters were collected: followers, following and favorites so the relationships with relevance metrics could be assessed.

We applied several statistical tools to characterize users in terms of the relevance metrics. We also implemented visualizations of different variables and the network for a better understanding of leading nodes characterization and topology.[1]

---

[1]The interactive dashboard with the data used for visualizations can be downloaded from `https://zenodo.org/record/3996654#.X0LLG9Mza3I`. The original Twitter data can be provided upon request.

Results

We compared the relevance in the network derived from the centrality metrics with the user' profile variables of Twitter: number of followers, number of following and retweet count. Figure 1 shows a scatter plots matrix among all variables. Principal diagonal of the figure shows the distribution of each variable which are normally characterized by a high concentration in low values and a very long tail of the distribution. These distributions imply that few nodes concentrate most part of the relevance within the network. More surprisingly, same distributions are observed for Twitter user' parameters such as number of followers or friends (following).

The scatter plots shows that there is no significant correlation between variables except for the pair betweenness and load centralities as it is expected because they have similar definitions. This fact is remarkable as different centrality metrics provide a different perspective of leading nodes within the network and it does not necessarily correlate with the amount of related users, but also in the content dynamics.

Users were ranked using on variable as the reference. Figure 3 summarizes the values for each descriptor of each leader after being ranked according to the eigenvalue centrality. Figure shows that even within the top ranked leaders there is a very large variability characterized by an exponential distribution for the eigenvalue parameter and very heterogeneous values for other relevance metrics or Twitter popularity metrics (followers, following, favorites and status count). The heterogeneous and unequal distribution suggested that a small number of nodes are powerful nodes within the network accumulating most part of the relevance and node connectivity as characterized by the eigenvalue metric. This fact requires further analysis to be interpreted as relevant nodes can be indeed social leaders in society or singular events of the network dynamics within the time window analyzed. Figure 1, 2 and 3 show that relevance may not be directly correlated with popularity or very high in Twitter. Figure 1 shows all possible histograms and scatters between variables describing the network of tweets (descriptors Table 1 and Table 2). The diagonal shows distributions that are normally very fast-decreasing distributions near zero except the closeness centrality that shows a quasi-symmetric curve. The scatter plots show that there is no significant correlation between variables which means that all descriptors convey relevant information to describe the network and the users. This emergent relevance of specific nodes is key to understand the propagation of information. Figure 2 shows scatter plots for ranked users based on eigenvalue centrality (the first 500 users are selected). This is a subset of Figure 1 and still shows that there are not correlation of variables in highly-ranked groups.

Figure 3 shows that there is a concentration of very few leading nodes in some variables because they are fat-tailed [26] . Only closeness shows a distribution that features high values more homogeneous for the ranked variables and therefore this variable is not fat-tailed.

Followers	Number of followers of the account
Following	Number of users that are followed by the account
Favorites	Number of favorites status (tweets) by other users
Status count	Number of tweets published by the account

Table 2: Descriptors of Twitter accounts

Figure 4 shows the ranking resulting from using the eigenvalue centrality as the reference. The values were saturated to the percentile 95 of the distribution to improve visualization and avoid the effect of single values with very out of range values. This visualization confirms the lack of correlation between variables and the highly asymmetric distribution of the descriptors. Figure 5 shows the ranking resulting from using current flow betweenness centrality as the reference. In this case, the distribution of this reference variable is smoother and shows a more gradual behavior of leaders. Of note, different centrality metrics lead to different a classification and characterization of nodes (Figure 4 vs Figure 5). Experimental work is required to understand how a specific centrality translates into a different information propagation pattern. For this purpose, we built a dashboard to browse the ranked nodes, their properties and relevance and also the network and its topology.

The occurrence of nodes with centrality values very far away from the distribution average is an important phenomenon when study social leaders. These nodes can play a role of information super-spreaders meaning that a few nodes transmit a lot of information or misinformation to other nodes. This can be appreciated specially in the eigenvalue centrality, whereas the current-flow betweenness centrality seems a more stable metric. This asymmetric distribution implies that there are powerful communities highly intra-connected whereas there are fewer nodes that serve as a bridge between communities.

A clear conclusion is that few nodes with high eigenvalue centrality have a lot of power in shaping the opinion and information within a community, that may or may not be closely distributed geographically, so they are clear influencers at least for a close group of people. However, nodes with high current-flow betweenness centrality are specially relevant to introduce relevant information into communities that, for instance, may have a negative narrative more indirectly. An open question is that these nodes have the necessary influence within the target community to propagate information and consolidate sentiment. A strategy could be to reinforce nodes with high current-flow betweenness and positive activity to be more influential and gain the necessary eigenvalue centrality within communities.

To assess how the nodes with high relevance are distributed we projected the network into graphs by selecting the subgraph of nodes with a certain level of relevance (threshold on the network). The resulting network graphs may not be therefore connected.

The eigenvalue-ranked graph shows high connectivity and very big nodes (see Fig. 6). This is consistent with the definition of eigenvalue centrality that highlights how a node is connected to nodes that are also highly connected. This structure has implications in the reinforcement of specific messages and information within high connected clusters which can act as promoters of solutions, sources of infor-

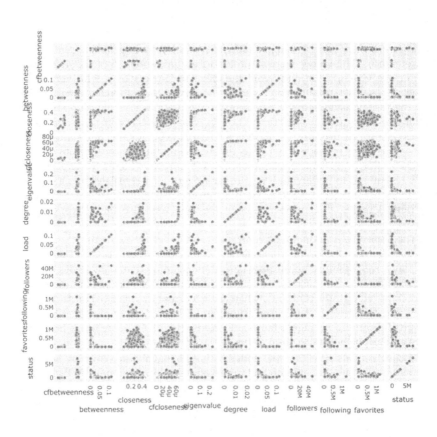

Figure 1: Matrix of histograms and scatter plots among all variables in log scale. Left to right and from the top to the bottom: current-flow betweenness, betweenness, closeness, current-flow closeness, eigenvalue, degree, load, followers, following, favorites and status. count

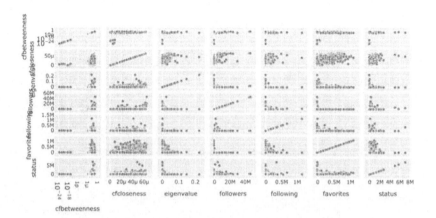

Figure 2: Matrix of histograms and scatter plots among all variables with logarithmic axis. It represents the correlations between the vars for the top 500 ranked users by eigenvalue. Left to right and from the top to the bottom: current-flow betweenness, current-flow closeness, eigenvalue, followers, following, favorites and status count.

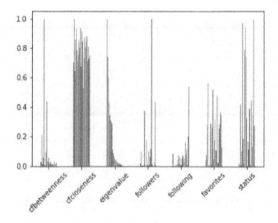

Figure 3: Distribution of the ranked users for each descriptor.

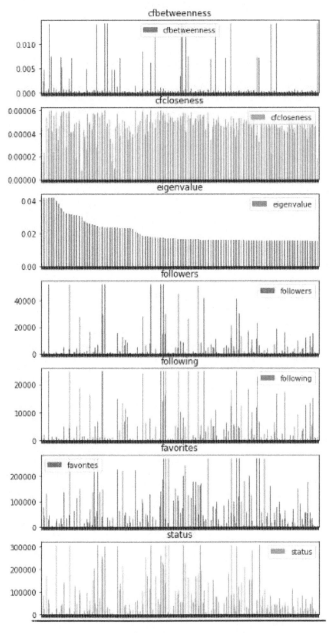

Figure 4: Mosaic of bar plots for ranked users according to eigenvalue centrality. Descriptors shown: current-flow betweenness, current-flow closeness, eigenvalue, followers, following, favorites and status count (the number of status in Twitter). Each bar for each user.

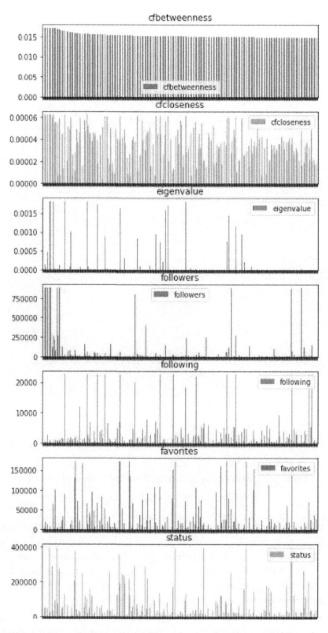

Figure 5: Mosaic of bar plots for ranked users according to current flow betweenness. Descriptors shown: current-flow betweenness, current-flow closeness, eigenvalue, followers, following, favorites and status count (the number of status in Twitter). Each bar for each user.

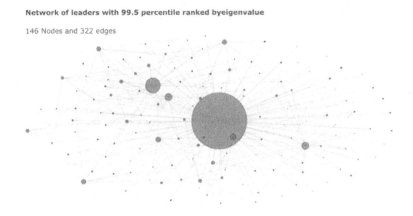

Figure 6: Graph of high-eigenvalue users.

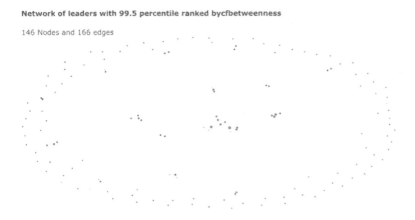

Figure 7: Graph of high-current flow betweenness users.

mation/misinformation or sentiment or may become lobbies. It is still remarkable that these nodes may not be those with more popularity according to Twitter metrics. It means that for given conversations and topics the dynamic network is not influenced by the popularity of the network. Further analysis is here required to understand the authority that emerges for specific topics based on the content of messages, dynamics of the network and the popularity.

The current flow betweenness shows an unconnected graph which is very interesting as decentralized nodes play a key role in transporting information through the network (see Fig. 7). This means that the connectors between communities and groups of opinion are distributed in the network and potentially geographically too. As mentioned, these nodes may not have high popularity or high connectivity (as measured by eigenvalue), but the messages they convey are transmitted across different communities. Further research is required to see the impact of these nodes

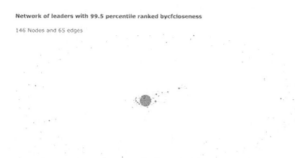

Figure 8: Graph of high-current flow closeness users.

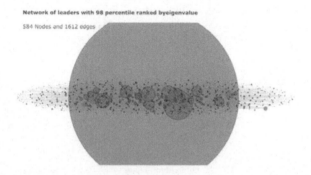

Figure 9: Graph of high-eigenvalue users (size 584 nodes).

into the narratives and the propagation of specific sentiments and topics into the communities these nodes interconnect as these nodes have great potential to build larger safety, well-informed and positive nets.

The current flow closeness shows also an unconnected graph which means that the social network is rather homogeneously distributed overall with parallel communities of information that do not necessarily interact with each other (see Fig. 8). These nodes, as the nodes characterized by current flow betweenness have great potential to interconnect communities as they are closer in the network to several communities at the same time. A research question is if the closeness would be sufficient to propagate and consolidate information and sentiment into the target communities.

By increasing the size of the graph (lowering the thresholds) more clusters can be observed, specially in the eigenvalue-ranked network which is consistent with the previous observations (Fig. 9-11). A super node may point out to a relevant institution or an anomaly in the network caused by a viral process or topic. The large connectivity of high eigenvalue centrality nodes may be also related to the size of the communities where few communities may be specially large and intra-connected concentrating the flows of information on a specific topic.

Some clusters also appear for the current flow betweenness and current flow closeness (see Fig.10, 11). These clusters may have a key role as highly relevant

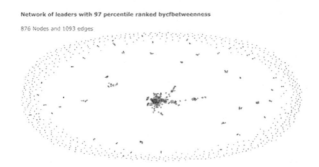

Figure 10: Graph of high-current flow betweenness users (size 876).

Figure 11: Graph of high-current flow closeness users (size 876).

hubs in establishing bridges between different communities of practice, knowledge or region-determined groups.

Discussion

Distributions of the centrality metrics indicate that there are some nodes with massive relevance. As the edges of the network are characterized in terms of flows between users, the relevance should be understood in terms of volume of information between communities or groups that are dynamically connected within a specific topic, in this case information related to COVID-19.

The relevant nodes are topological events within the flow of communication through the network [68] that require further contextualization to be interpreted. These nodes can propagate misinformation or make news or messages viral in different ways and with different network length scopes depending on the type of centrality that characterizes them. High eigenvalue nodes will do dense propagation within communities whereas current flow closeness and betweenness will do sparser and widely spread propagation of information. Experimental work is required to optimize the necessary balance between these types of centrality to properly and effectively propagate good information and positive sentiment through the network as an opposition to infodemics. Further research is required to understand the cause of this massive relevance events, for instance, if it is related to a relevant concept or message or whether it is an emerging event of the network dynamics and topology. Another way to assess these nodes is if they are consistently behaving this way along time or they are a temporal event. Also, it may be necessary to contextualize with the type of content they normally spread to understand their exceptional relevance.

Besides the existence of massive relevance nodes, the quantification and understanding of the distribution of high relevant nodes has a lot of potential applications to spread messages to reach a wide number of users within the network. This is important for inclusiveness of information and to target all types of communities including those that are more vulnerable and can be more affected by infodemics and real impact of the pandemic, both in epidemiological and socio-economic terms. Current flow betweenness particularly seems a good indicator to identify nodes to create a safety net in terms of information and positive messages. The distribution of the nodes could be approached for the general network or for different layers or subnetworks, isolated depending on several factors: type of interaction, type of content or some other behavioral pattern.

We have also developed a first version of an interactive graph visualization to browse the relevance of the network and dynamically investigate how relevant nodes are connected and how specific parts of the graph are ranked to really understand the distribution of the relevance variables.

5 Conclusions

Governance is an emergent opportunity to develop more sustainable, resilient, anti-fragile and ethical societies. However, a simplified version of data-driven and evidence-based mechanisms can lead to very negative outcomes and also generate distrust in the population against technology and computation-driven decisions.

When considering complex and systemic problems such as pandemics, holistic approaches integrating Science and Collective Intelligence should prevail over linear decision making. The design of computation-powered mechanisms should promote both depth and wideness of perspectives to improve judgement and wisdom of population, teams and boards rather than instrumental tools to justify the same decision making processes that cannot cope with problems such as the COVID-19 pandemic. Current AI approaches may provide certain depth in data analysis, but they lack the wideness, which may lead into decisions with uncontrolled impact. This pandemic is an opportunity to define what is a good computation and data-driven decision making to build a more resilient and sustainable future.

The human side of governance and policy making cannot be overlooked. The pandemic is challenging current organizations of all kind around the globe, from governments to private corporations to academia. A collaborative effort in academia and the scientific community has led to a rapid response in understanding COVID-19 better [51]. However, these efforts have to be integrated into governance platforms with capacities and capabilities to propose deep transformations and new mechanisms that are required for response to systemic threats. Technology should be the basis for activation, data sharing, collaboration, exchanges and also mediate in partnerships to catalyze responsible action. Intra-organization and inter-organization relationships management is a unique opportunity for a world that is interconnected and has computational frameworks to help decision.

Another key issue is the lack of leadership, both in terms of action and also generating narratives that help the society to face the pandemic. The unbearable responsibility for many decision makers and politicians of causing harm (socio-economic, cultural, spiritual and health) through strict policies and disruptive measures is likely causing inaction in many local and national governments. We must open the chance for computational governance to complement existing mechanisms and allow disruption at scale.

Computational platforms have also an intrinsic risk to scale up negative outcomes, biases and misinformation. The future of digitalization and evidence-based likely requires designing systems that regulate the ethical use of technologies, not only in a personal way (i.e. regarding privacy), but also the systemic and behavioral dimension of technology [76].

Collaboration and coordination should be the ground for governance. However, these platforms need to introduce disruptive elements based on organizational and technological innovation. Societies and governments will need accelerators and catalyzers to tackle the upcoming systemic challenges in the next decade to recover from COVID-19 and meet the 2030 Agenda, the United Nations commitment and roadmap for sustainable development . Collective efforts and truth-driven narratives must be promoted to fight polarization and align efforts in certain challenges that will affect us all. Necessarily, this process must be undertaken laterally and vertically considering the local problems of people but incrementing the awareness of the systemic problems we face and the need to overcome selfishness and hatred. Minorities can lead to unfair decisions if there is a lack of proper integrative governance and ethical frameworks. It is now only acceptable to achieve these goals in an inclusive way, so acceleration and inclusiveness must be coupled into the actions for new policies and governance platforms. This is not only an ethical principle, but a design, implementation and deployment principle for an interconnected world that is resilient and sustainable. Ethical principles that promote protection, action and future projection should be the basis of evidence-based systems to not become empty computational boxes. This is required to avoid self-complexity of the environment to be more dominant that the actual society and their decisions. This problem has been discussed within Ashby's Law.

Linear decision making involves several types of models such as SIR models or agent-based epidemiological models. These models are normally used as predictive systems but the model does not change the decision making process itself. It is necessary to rethink decision making process as Collective Intelligence problems that demand new algorithms and network-like topologies to make decisions that are supported by leaders in different aspects. Multi-level leadership is important for multi-scale analysis and governance. The analysis presented is an step forwards in unveiling influence networks that promote robustness and resilience rather than power control. Through the propagation of sentiment and information is possible not only to characterize the topology of the network, but also the meaning and the topography of the network. By calibrating weights of the nodes of the network it would be possible to reconfigure them for a positive and constructive global dialogue for building narratives.

Most likely, designing such a societal system will be a process that requires experimentation and space for failure demanding also robustness of the system to mitigate negative outcomes and impacts. This is a good approach for computational frameworks rather than feeding linear decision making processes which may turn into biased and oversimplified computational intelligence that may spoil the potential of computational and data to empower a more resilient and thriving society.

Bibliography

[1] M. Salathe, L. Bengtsson, T. J. Bodnar, D. D. Brewer, J. S. Brownstein, C. Buckee, E. M. Campbell, C. Cattuto, S. Khandelwal, and P. L. Mabry, "Digital epidemiology," *PLoS computational biology*, vol. 8, no. 7, p. e1002616, 2012.

[2] M. Boguná, R. Pastor-Satorras, and A. Vespignani, "Absence of epidemic threshold in scale-free networks with degree correlations," *Physical review letters*, vol. 90, no. 2, p. 028701, 2003.

[3] D. Balcan, B. Gonçalves, H. Hu, J. J. Ramasco, V. Colizza, and A. Vespignani, "Modeling the spatial spread of infectious diseases: The global epidemic and mobility computational model," *Journal of computational science*, vol. 1, no. 3, pp. 132–145, 2010.

[4] S. Meloni, N. Perra, A. Arenas, S. Gómez, Y. Moreno, and A. Vespignani, "Modeling human mobility responses to the large-scale spreading of infectious diseases," *Scientific reports*, vol. 1, p. 62, 2011.

[5] Y. Moreno and A. Vazquez, "Disease spreading in structured scale-free networks," *The European Physical Journal B-Condensed Matter and Complex Systems*, vol. 31, no. 2, pp. 265–271, 2003.

[6] D. Pastor-Escuredo and E. Frias-Martinez, "Flow descriptors of human mobility networks," *arXiv preprint arXiv:2003.07279*, 2020.

[7] A. Wesolowski, N. Eagle, A. J. Tatem, D. L. Smith, A. M. Noor, R. W. Snow, and C. O. Buckee, "Quantifying the impact of human mobility on malaria," *Science*, vol. 338, no. 6104, pp. 267–270, 2012.

[8] A. Wesolowski, C. O. Buckee, K. Engø-Monsen, and C. J. E. Metcalf, "Connecting mobility to infectious diseases: the promise and limits of mobile phone data," *The Journal of infectious diseases*, vol. 214, no. 4, pp. S414–S420, 2016.

[9] C. Lynch and C. Roper, "The transit phase of migration: circulation of malaria and its multidrug-resistant forms in africa," *PLoS medicine*, vol. 8, no. 5, 2011.

[10] C. Jackson, P. Mangtani, J. Hawker, B. Olowokure, and E. Vynnycky, "The effects of school closures on influenza outbreaks and pandemics: systematic review of simulation studies," *PLoS one*, vol. 9, no. 5, p. e97297, 2014.

[11] S. B. Nafisah, A. H. Alamery, A. Al Nafesa, B. Aleid, and N. A. Brazanji, "School closure during novel influenza: a systematic review," *Journal of Infection and Public Health*, vol. 11, no. 5, pp. 657–661, 2018.

[12] B. J. Cowling, S. T. Ali, T. W. Ng, T. K. Tsang, J. C. Li, M. W. Fong, Q. Liao, M. Y. Kwan, S. L. Lee, and S. S. Chiu, "Impact assessment of non-pharmaceutical interventions against coronavirus disease 2019 and influenza in hong kong: an observational study," *The Lancet Public Health*, 2020.

[13] A. de Niet, B. L. Waanders, and I. Walraven, "The role of children in the transmission of mild sars-cov-2 infection," *Acta Paediatrica*, 2020.

[14] P. Zimmermann and N. Curtis, "Coronavirus infections in children including covid-19: an overview of the epidemiology, clinical features, diagnosis, treatment and prevention options in children," *The Pediatric infectious disease journal*, vol. 39, no. 5, p. 355, 2020.

[15] J. F. Ludvigsson, "Systematic review of covid-19 in children shows milder cases and a better prognosis than adults," *Acta Paediatrica*, vol. 109, no. 6, pp. 1088–1095, 2020.

[16] L. Meyers, "Contact network epidemiology: Bond percolation applied to infectious disease prediction and control," *Bulletin of the American Mathematical Society*, vol. 44, no. 1, pp. 63–86, 2007.

[17] K. Prem, A. R. Cook, and M. Jit, "Projecting social contact matrices in 152 countries using contact surveys and demographic data," *PLoS computational biology*, vol. 13, no. 9, p. e1005697, 2017.

[18] D. Martín-Calvo, A. Aleta, A. Pentland, Y. Moreno, and E. Moro, "Effectiveness of social distancing strategies for protecting a community from a pandemic with a data driven contact network based on census and real-world mobility data," report, Working paper, https://covid-19-sds. github. io (accessed April 18, 2020), 2020.

[19] J. Zhang, M. Litvinova, Y. Liang, Y. Wang, W. Wang, S. Zhao, Q. Wu, S. Merler, C. Viboud, and A. Vespignani, "Changes in contact patterns shape the dynamics of the covid-19 outbreak in china," *Science*, 2020.

[20] Q. Bi, Y. Wu, S. Mei, C. Ye, X. Zou, Z. Zhang, X. Liu, L. Wei, S. A. Truelove, and T. Zhang, "Epidemiology and transmission of covid-19 in 391 cases and 1286 of their close contacts in shenzhen, china: a retrospective cohort study," *The Lancet Infectious Diseases*, 2020.

[21] S. K. Brooks, L. E. Smith, R. K. Webster, D. Weston, L. Woodland, I. Hall, and G. J. Rubin, "The impact of unplanned school closure on children's social contact: rapid evidence review," *Eurosurveillance*, vol. 25, no. 13, p. 2000188, 2020.

[22] A. Aleta, D. Martín-Corral, A. P. y Piontti, M. Ajelli, M. Litvinova, M. Chinazzi, N. E. Dean, M. E. Halloran, I. M. Longini Jr, and S. Merler, "Modelling the impact of testing, contact tracing and household quarantine on second waves of covid-19," *Nature Human Behaviour*, pp. 1–8, 2020.

[23] K. Prem, K. van Zandvoort, P. Klepac, R. M. Eggo, N. G. Davies, A. R. Cook, and M. Jit, "Projecting contact matrices in 177 geographical regions: an update and comparison with empirical data for the covid-19 era," *medRxiv*, 2020.

[24] L. Willem, S. Abrams, O. Petrof, P. Coletti, E. Kuylen, P. Libin, S. Mogelmose, J. Wambua, S. A. Herzog, and C. Faes, "The impact of contact tracing and household bubbles on deconfinement strategies for covid-19: an individual-based modelling study," *medRxiv*, 2020.

[25] J. Norman, Y. Bar-Yam, and N. N. Taleb, "Systemic risk of pandemic via novel pathogens—coronavirus: A note," *New England Complex Systems Institute (January 26, 2020)*, 2020.

[26] N. N. Taleb, "Statistical consequences of fat tails: Real world preasymptotics, epistemology, and applications," 2019.

[27] J. Bullock, K. H. Pham, C. S. N. Lam, and M. Luengo-Oroz, "Mapping the landscape of artificial intelligence applications against covid-19," *arXiv preprint arXiv:2003.11336*, 2020.

[28] W. Naude and R. Vinuesa, "Data, global development, and covid-19: Lessons and consequences," report, World Institute for Development Economic Research (UNU-WIDER), 2020.

[29] K. Shu, A. Sliva, S. Wang, J. Tang, and H. Liu, "Fake news detection on social media: A data mining perspective," *ACM SIGKDD Explorations Newsletter*, vol. 19, no. 1, pp. 22–36, 2017.

[30] D. M. Lazer, M. A. Baum, Y. Benkler, A. J. Berinsky, K. M. Greenhill, F. Menczer, M. J. Metzger, B. Nyhan, G. Pennycook, and D. Rothschild, "The science of fake news," *Science*, vol. 359, no. 6380, pp. 1094–1096, 2018.

[31] V. Bakir and A. McStay, "Fake news and the economy of emotions: Problems, causes, solutions," *Digital journalism*, vol. 6, no. 2, pp. 154–175, 2018.

[32] H. Allcott and M. Gentzkow, "Social media and fake news in the 2016 election," *Journal of economic perspectives*, vol. 31, no. 2, pp. 211–36, 2017.

[33] J. Zarocostas, "How to fight an infodemic," *The Lancet*, vol. 395, no. 10225, p. 676, 2020.

[34] M. A. Peters, P. Jandrić, and P. McLaren, "Viral modernity? epidemics, infodemics, and the 'bioinformational' paradigm," 2020.

[35] A. Morales, J. Borondo, J. C. Losada, and R. M. Benito, "Measuring political polarization: Twitter shows the two sides of venezuela," *Chaos: An Interdisciplinary Journal of Nonlinear Science*, vol. 25, no. 3, p. 033114, 2015.

[36] C. Castellano and R. Pastor-Satorras, "Thresholds for epidemic spreading in networks," *Physical review letters*, vol. 105, no. 21, p. 218701, 2010.

[37] R. Vinuesa, A. Theodorou, M. Battaglini, and V. Dignum, "A socio-technical framework for digital contact tracing," *arXiv preprint arXiv:2005.08370*, 2020.

[38] S. Altmann, L. Milsom, H. Zillessen, R. Blasone, F. Gerdon, R. Bach, F. Kreuter, D. Nosenzo, S. Toussaert, and J. Abeler, "Acceptability of app-based contact tracing for covid-19: Cross-country survey evidence," *medRxiv*, 2020.

[39] G. Berman, K. Carter, M. G. Herranz, and V. Sekara, "Digital contact tracing and surveillance during covid-19. general and child-specific ethical issues," report, 2020.

[40] T. W. Malone, "How human-computer'superminds' are redefining the future of work," *MIT Sloan Management Review*, vol. 59, no. 4, pp. 34–41, 2018.

[41] P. Lévy and R. Bononno, *Collective intelligence: Mankind's emerging world in cyberspace*. Perseus books, 1997.

[42] G. Mulgan, "Artificial intelligence and collective intelligence: the emergence of a new field," *AI SOCIETY*, vol. 33, no. 4, pp. 631–632, 2018.

[43] J. Boy, D. Pastor-Escuredo, D. Macguire, R. M. Jimenez, and M. Luengo-Oroz, *Towards an understanding of refugee segregation, isolation, homophily and ultimately integration in Turkey using call detail records*, pp. 141–164. Springer, 2019.

[44] A. J. Morales, X. Dong, Y. Bar-Yam, and A. 'Sandy'Pentland, "Segregation and polarization in urban areas," *Royal Society Open Science*, vol. 6, no. 10, p. 190573, 2019.

[45] D. Pastor-Escuredo and C. Tarazona, "Characterizing information leaders in twitter during covid-19 crisis," *arXiv preprint arXiv:2005.07266*, 2020.

[46] M. Mazzucato, R. Kattel, and J. Ryan-Collins, "Challenge-driven innovation policy: towards a new policy toolkit," *Journal of Industry, Competition and Trade*, vol. 20, no. 2, pp. 421–437, 2020.

[47] H. Cho, D. Ippolito, and Y. W. Yu, "Contact tracing mobile apps for covid-19: Privacy considerations and related trade-offs," *arXiv preprint arXiv:2003.11511*, 2020.

[48] Q. Tang, "Privacy-preserving contact tracing: current solutions and open questions," *arXiv preprint arXiv:2004.06818*, 2020.

[49] D. Pastor-Escuredo and R. Vinuesa, "Towards and ethical framework in the complex digital era," *arXiv preprint arXiv:2010.10028*, 2020.

[50] J. L. J. G. David Pastor-Escuredo, Gianni Giacomelli, "Ethical and sustainable future of work," *DIECISIETE*, 2020 (In press).

[51] M. Luengo-Oroz, K. H. Pham, J. Bullock, R. Kirkpatrick, A. Luccioni, S. Rubel, C. Wachholz, M. Chakchouk, P. Biggs, and T. Nguyen, "Artificial intelligence cooperation to support the global response to covid-19," *Nature Machine Intelligence*, pp. 1–3, 2020.

[52] D. Pastor-Escuredo, Y. Torres, M. Martínez-Torres, and P. J. Zufiria, "Rapid multi-dimensional impact assessment of floods," *Sustainability*, vol. 12, no. 10, p. 4246, 2020.

[53] M. Cinelli, W. Quattrociocchi, A. Galeazzi, C. M. Valensise, E. Brugnoli, A. L. Schmidt, P. Zola, F. Zollo, and A. Scala, "The covid-19 social media infodemic," *arXiv preprint arXiv:2003.05004*, 2020.

[54] J. Hua and R. Shaw, "Corona virus (covid-19)"infodemic" and emerging issues through a data lens: The case of china," *International journal of environmental research and public health*, vol. 17, no. 7, p. 2309, 2020.

[55] R. J. Medford, S. N. Saleh, A. Sumarsono, T. M. Perl, and C. U. Lehmann, "An" infodemic": Leveraging high-volume twitter data to understand public sentiment for the covid-19 outbreak," *medRxiv*, 2020.

[56] A. Vaezi and S. H. Javanmard, "Infodemic and risk communication in the era of cov-19," *Advanced Biomedical Research*, vol. 9, 2020.

[57] L. G. Militello, E. S. Patterson, L. Bowman, and R. Wears, "Information flow during crisis management: challenges to coordination in the emergency operations center," *Cognition, Technology Work*, vol. 9, no. 1, pp. 25–31, 2007.

[58] F. Greenwood, C. Howarth, D. Escudero Poole, N. A. Raymond, and D. P. Scarnecchia, "The signal code: A human rights approach to information during crisis," *Harvard, MA*, 2017.

[59] L. Gao, C. Song, Z. Gao, A.-L. Barabási, J. P. Bagrow, and D. Wang, "Quantifying information flow during emergencies," *Scientific reports*, vol. 4, p. 3997, 2014.

[60] F. Pierri and S. Ceri, "False news on social media: A data-driven survey," *ACM SIGMOD Record*, vol. 48, no. 2, pp. 18–27, 2019.

[61] S. MacAvaney, H.-R. Yao, E. Yang, K. Russell, N. Goharian, and O. Frieder, "Hate speech detection: Challenges and solutions," *PloS one*, vol. 14, no. 8, 2019.

[62] B. Ghanem, P. Rosso, and F. Rangel, "An emotional analysis of false information in social media and news articles," *ACM Transactions on Internet Technology (TOIT)*, vol. 20, no. 2, pp. 1–18, 2020.

[63] K. Popat, S. Mukherjee, A. Yates, and G. Weikum, "Declare: Debunking fake news and false claims using evidence-aware deep learning," *arXiv preprint arXiv:1809.06416*, 2018.

[64] N. Ruchansky, S. Seo, and Y. Liu, "Csi: A hybrid deep model for fake news detection," in *Proceedings of the 2017 ACM on Conference on Information and Knowledge Management*, pp. 797–806.

[65] S. Singhania, N. Fernandez, and S. Rao, "3han: A deep neural network for fake news detection," in *International Conference on Neural Information Processing*, pp. 572–581, Springer.

[66] G. Miritello, E. Moro, and R. Lara, "Dynamical strength of social ties in information spreading," *Physical Review E*, vol. 83, no. 4, p. 045102, 2011.

[67] J. L. Iribarren and E. Moro, "Impact of human activity patterns on the dynamics of information diffusion," *Physical review letters*, vol. 103, no. 3, p. 038702, 2009.

[68] A. J. Morales, J. Borondo, J. C. Losada, and R. M. Benito, "Efficiency of human activity on information spreading on twitter," *Social Networks*, vol. 39, pp. 1–11, 2014.

[69] J. Borondo, A. Morales, R. Benito, and J. Losada, "Multiple leaders on a multilayer social media," *Chaos, Solitons Fractals*, vol. 72, pp. 90–98, 2015.

[70] P. Balkundi and M. Kilduff, "The ties that lead: A social network approach to leadership," *The leadership quarterly*, vol. 17, no. 4, pp. 419–439, 2006.

[71] F. Bodendorf and C. Kaiser, "Detecting opinion leaders and trends in online social networks," in *Proceedings of the 2nd ACM workshop on Social web search and mining*, pp. 65–68.

[72] A. De Brún and E. McAuliffe, "Exploring the potential for collective leadership in a newly established hospital network," *Journal of Health Organization and Management*, 2020.

[73] K. Fransen, S. Van Puyenbroeck, T. M. Loughead, N. Vanbeselaere, B. De Cuyper, G. V. Broek, and F. Boen, "Who takes the lead? social network analysis as a pioneering tool to investigate shared leadership within sports teams," *Social networks*, vol. 43, pp. 28–38, 2015.

[74] A. Goyal, F. Bonchi, and L. V. Lakshmanan, "Discovering leaders from community actions," in *Proceedings of the 17th ACM conference on Information and knowledge management*, pp. 499–508.

[75] E. Iakhnis and A. Badawy, "Networks of power: Analyzing world leaders interactions on social media," *arXiv preprint arXiv:1907.11283*, 2019.

[76] I. Rahwan, M. Cebrian, N. Obradovich, J. Bongard, J.-F. Bonnefon, C. Breazeal, J. W. Crandall, N. A. Christakis, I. D. Couzin, and M. O. Jackson, "Machine behaviour," *Nature*, vol. 568, no. 7753, pp. 477–486, 2019.

The Future of Work and Forced Automation after COVID-19

Jose Balsa-Barreiro* and Elia Rossi

MIT Media Lab, Cambridge, MA, USA.
*Corresponding author: jobalbar@mit.edu

The global labor market has been radically changing over the last decades. The ongoing globalization process has implied that most of the manufacturing sectors in western countries have been relocated. For some years, we have been going through a Technological Revolution with significant consequences on the economies of many countries. According to the International Labor Organization, around 37.5 per cent of the total world workforce are subject to vulnerable employment and this number is expected to increase in the upcoming decade. In addition, the emergence of the COVID-19 pandemic is strongly affecting the job market, ranging from remote working to the need to set up reliable and virus-free supply chains in companies and businesses. Although the landscape after the pandemic is still very volatile, many companies are already shifting towards forced automation in their production. In this paper, we consider these factors and their implications to draw potential scenarios of the future labor market on a global scale. Keywords: 4th Industrial Revolution; Artificial Intelligence; automation; COVID-19 pandemic; decent work; Machine Learning; future labor market; future scenarios; employment prospects; robotics; technological revolution

1 Introduction

Recent technological developments have undoubtedly had an impact on many aspects of current societies. One of the most important is how these will influence our working life. In recent years, a large number of experts, institutions, governments and policymakers have shown possible scenarios about the future of work. They did so within the context of relevant technological shifts brought about by the so-called 4th Industrial Revolution. However, the sudden emergence of the COVID-19 pandemic has appeared as a shock in our lives creating great uncertainty. Among other aspects, we must rethink what new scenarios can be envisaged on the future of work. Today, more than ever, it makes sense to analyze changes that may occur at the intersection between the 4th Industrial Revolution and the COVID-19 pandemic by pointing out what implications are expected to occur in the global labor market. To do this, we will analyze the trends recently observed for understanding the magnitude of changes and the directions in which these point. We will review some potential scenarios provided by experts within a very volatile political and socioeconomic context on a global scale.

The sudden emergence of the COVID-19 pandemic has been negatively influencing individuals and societies as a whole. To this day, one of the major uncertainties is related to the economic impact of this pandemic. Many academics and experts are trying to understand this impact by proposing a large number of potential scenarios of economic regeneration (Ortega 2020). However, as the health crisis has not finished, it is still hazardous to predict the future. It will depend on multiple factors such as the new geopolitical relations between the western countries and China, the political understanding between Europe and the United States, and the policies for reconstruction and economic revitalization. What does already look clear is that the global labor market will undergo profound changes due to the existing inertia experienced in previous years and also due to the impact of the pandemic.

2   Analysis and Discussion

Before the COVID-19 emergency, the global labor market was already going through radical changes over the preceding decades due to two main factors: (a) the process of industrial relocation in western countries, and (b) the so-called Technological Revolution. The United States is the great paradigm behind these dynamics. In the mid-20th century, those states located in the center of the country (aka flying states) based their economies on manufacturing tangible assets. In 1950, around 35 percent of Americans were working in the manufacturing sector, while in some of these flying states this rate was over 50 per cent (Kozmetsky and Yue 2005). In the last 30 years, most American manufacturing industry has been outsourced to third world (developing) countries where companies sourced reduced production costs, much lower requirements for labor protection, and minimal environmental regulations, among other advantages. In consequence, the service sector has come to dominate the US economy, accounting for 80 per cent of the employment and 77 percent of national GDP in 2019 (World Bank 2020). From a geographical perspective, it has led to the agglomeration of people in cities and the reinforcement of urban economies. Most job opportunities are to be found in cities and people migrate there. In addition, of course, this helps with understanding the increasing spatial polarization between urban and rural areas in the US, but also elsewhere (Hedayatifar et al. 2019). In fact, these same dynamics have been replicated in the rest of the western countries. This is the starting point for better understanding most of the social and political dynamics that these countries have been experiencing in the last few years.

There is a second major turning point: the subprime crisis in 2008. After that year, most western countries have faced a qualitative degradation in employment under the pretext of the global economic crisis (Balsa-Barreiro, 2013). However, this employment degradation could also show an anticipation for a clear drop in the number of employees due to the average robot prices having fallen more than labor costs, which encouraged companies to shift towards fast automation in many countries (Tilley 2017). Optimization is associated with fragility (Taleb 2012), and fragility means lack of adaptation and less chances of survival (which is critical for operating in uncertainty). Supply chains can be automated, but if they are not restructured, they continue to be fragile to massive disruption (Balsa, Vie, Morales, Cebrian 2020).

Before the emergence of COVID-19, the International Labor Organization (ILO) estimated that 1,400 million workers in the world (around 37.5 per cent of the total) were working in vulnerable employment conditions (ILO 2018). This rate is likely to be increased during the pandemic. Future prospects are unpromising due to the advances and challenges generated by technologies related to Artificial Intelligence (AI), Machine Learning (ML), Robotics, and Industrial Automation. Such is the magnitude of the upcoming changes that some experts refer to this age as the 4th Industrial Revolution (Schwab 2016).

At this point, the role of technology has become self-contradictory. Even if technology has been created by and for supporting humankind, humans are afraid of its impact on the labor market. Previous industrial revolutions have shown how this concern has always existed behind luddism, a reactionary movement against the drastic changes resulting from any kind of technological revolution (Klein 2019; Conniff 2011). However, this collective hysteria concerning technological shifts may be due to people's distrust in who is the driver of the changes. This is happening right now, where the first question is if we are really facing a disruptive scenario (Morgan 2020).

Several factors intervene in our collective fear of technology. According to IFR, the operational stock of industrial robots was approximately 2,700 million robots on a global scale, a number which has grown exponentially in recent years. This number will continue its exponential growth in the coming years because it is quite likely that COVID-19 will accelerate the industrial processes of automation and digital transformation (Dell Technologies 2020). In fact, this seems to be a natural post-crisis trend of any sort, where companies try to increase production efficiency while reducing related labor costs. This inevitably leads to accelerated automation, starting with those jobs which are easier to replace such as those undertaken by blue-collar workers. Jaimovich and Siu (2012) demonstrated that jobs which are focused mainly on routine and repetitive tasks were the first to have been automated after the 2008 financial crisis. Companies like Philips Electronics or Foxconn, the largest private employer in Taiwan, conducted a new wave of automation in both manufacturing and distribution right after the crisis (Markoff 2012). However, unexpected things can happen when large scale technologies are introduced in the population in a top down manner, such as second order consequences and hidden risks (Taleb 2012).

In addition, most of these jobs never returned. Furthermore, because of the impact of COVID-19 and the threat of future pandemics, many companies have been forced to organize supply chains that are safe and free from viruses, whereas other businesses have replaced humans with machinery to avoid workplace infections and to keep operating costs low (Ding and Saenz 2020; Semuels 2020). This is mandatorily requiring more automation and less human contact. Just as an example, some companies are already investing in deployments of swarm robots, which present important functional benefits (Luca et al. 2019).

With part of the world with lockdowns and mobility restrictions, the pandemic is also increasing social inequalities due to a disparate impact of the virus on different social classes and types of businesses. Thus, while restaurants and airports suffer a very severe drop in their activities, other productive sectors keep working remotely. Frey and Osborne (2013) argue that high-income earners are five times more likely to work remotely. Before the pandemic, the most vulnerable jobs with a higher risk of being automated were precisely those which were carried out by low-educated

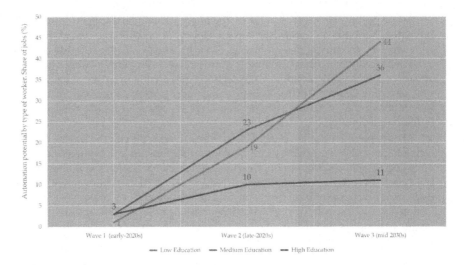

Figure 1: Automation potential by type of worker for the three waves of automation that might unfold over the period 2020-2035. The X-axis represents time range/waves and the Y-axis represents the share of jobs. (Data source: PwC estimates based on analysis of OECD PIAAC data 2017).

workers (Figure 1). The virus, being more a threat to certain types of jobs (Lu 2020, Shendruk 2020), reinforces the same previous trend, especially threatening the most vulnerable groups (Figure 2).

Nevertheless, the ongoing technological revolution and the impact of the pandemic might radically change the work prospects for white-collar employees too. With this new wave of forced remote working, some of the tech companies in Silicon Valley (such as Twitter, Square, Coinbase, Box, Shopify or Facebook) have already decided to shift to permanent homeoffice working even after the pandemic. Meanwhile, other companies are considering this shift depending on whether it maintains high levels of productivity. Remote working allows companies to reduce office costs, cut down on travel expenses (and time), and also avoids the geographical restriction of hiring employees with lower salary expectations. Just a few months ago, Facebook announced that half of its 48,000 employees will permanently work remotely by 2030 (Price 2020). Of course, these changes will have a vast impact in the urban economies of the regions where these companies are located by affecting aspects such as real estate, transportation, leisure, and social diversity, among others.

The 4th Industrial Revolution goes beyond just simple task automation and a shift to remote working. Some recent examples illustrate this statement. In 2017, two doctoral students at MIT presented a very cost-efficient system of school buses in Boston, United States. Their proposal was based on an algorithm that eliminated 75 bus routes and enabled them to save up to USD 5 million per year (McGinty 2017). In 2019, the 178-year-old British tour company with 22,000 employees, Thomas Cook, collapsed. In essence, this collapse is a consequence of the radical transformation which the travel industry and online booking services have experienced over the last number of years (Holton and Faulconbridge 2019). Also, the emergence of electric

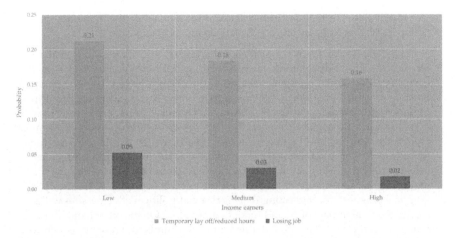

Figure 2: Risk probability (measured from 0 to 1) of temporary lay off/reduced hours and losing job by income group in the European Union in Q2 2020. (Data source: Eurostat 2020).

self-driving trucks could potentially disrupt the sector of truck drivers, which is currently employing around 3.5 million workers in the US alone(ATA 2017).

At first glance, these examples suggest that technology is leading to a clear reduction in the number of jobs (Watson 2017). Nevertheless, this approach must be nuanced. The 4th Industrial Revolution will eliminate a great number of jobs, especially in certain sectors, but it will also open new labor opportunities. It can be clearly observed in the Big Tech companies. The company Amazon, with around 700,000 employees worldwide, has recently announced 30,000 new jobs in the US, ranging from engineers to freight specialists (Weise 2019). This growth is magnified during the COVID-19 pandemic due to most shopping activities having been forwarded to digital platforms. Nevertheless, although the common perception of working in tech and innovative companies is bright, it is not always the case. Companies like Uber or Lyft, which are already threatening the jobs of millions of taxi drivers, are under investigation in Europe because of the vulnerable working conditions of their employees.

Since before the COVID-19 pandemic, there has been a discussion regarding how the current Tech Revolution will affect employment. Just a few years ago, consulting firms such as PwC (2018) and Gartner (2017) concluded that by 2020 the AI sector would create as many jobs as it eradicated, even showing a slightly positive trend. The current technological revolution shows relevant differences in comparison to previous industrial revolutions. Traditionally, changes derived from them affected only employees in charge of routine tasks, i.e. the so-called blue-collar workers. However, the recent impact of AI is also threatening those employees that carry out highly skilled tasks, i.e. the so-called white-collar workers. Just as an example, 600 traders of Goldman Sachs in New York headquarters were replaced by trading machinery that are supported by 200 computer engineers (Byrnes 2017). Nedelkoska and Quintini (2018) found nearly half of jobs are vulnerable to automation. According to Gartner (2017) and PwC (2018), the healthcare and educational sector will show an increase in the demand for jobs while the manufacturing and transport sectors will concentrate the largest losses of jobs.

The Pew Research Center analyzed the impact and implications that AI and robotics would produce on the labor market by 2025 (Smith and Anderson 2014). Obviously, they did it without considering the effects derived from the pandemic. For this purpose, they asked 1,896 experts whose answers were clustered in two groups: (a) those who expected a positive or neutral impact, and (b) those who expected a mostly negative impact. The first group or tech-optimists (52%) mostly argued that automation would create new needs, which would increase demand for new jobs. This hypothesis would explain why the most advanced regions (and urban areas) always demand more workers by concentrating the largest share of regional wealth. We must remark that some experts included in this group kept a skeptical perspective, suggesting that legal/social/political/ethical issues will surely reduce the final impact of AI and robotics in real-life. On the other hand, the second group or tech-pessimists (48%) defined an upcoming labor scenario, which would be unsustainable for most people due to a very volatile and ever-changing labor market. It would lead to the destruction and progressive decrease of life quality of the middle classes. Some of these theorists anticipated a post-work scenario where robots might mostly replace the human workforce. Among the consequences, the end of the traditional labor structure and the need to redefine the proper concept of work (Rifkin, 1995).

It is expected that the ongoing technological revolution, together with the entire crisis caused by the COVID-19, might reduce regulated and decent employment. In this scenario, the commitment between employer and employee would be reduced to its minimal expression. According to supporters of these sorts of policies, this would boost labor market efficiency, while the critics argue that employees would be totally unprotected in their jobs. In addition, with the pace of technological shifts, employees' skills will probably change faster over time, resulting in a group of workers whose skills become constantly outdated. Given these projections, Harari (2018) predicts the emergence of a massive and new unworking class referring to those people devoid of any economic, political or even artistic value that could contribute neither to the job market nor to society.

Some governments are trying to implement strategies to face any sort of negative scenarios which might emerge. Probably, the most controversial countermeasure is the universal basic income (UBI). Although this measure was mostly supported by leftist ideologies in the past, nowadays it seems different. Take for example, the experiment called Y-Combinator which has shown how some tech companies are open to the actual application of this political action (Winick 2018). Some experts warn about the negative implications related to the UBI as the weakening of the welfare state in some countries or the emergence of dual-class societies where the big corporations would have a huge control and excessive power over working classes. Therefore, although the UBI could reduce extreme poverty across the globe, it could paradoxically increase inequalities between social classes. However, when interacting with complex systems as human populations it is important to maintain precaution because the simplification of reality required for decision making can omit relevant information and create further harm. See the discussion about iatrogenics in Taleb 2012. Naive interventionism to improve things that leads to major disruption and creates new problems.

Fortunately, some authors like Manson (2015) share an optimistic perspective of the future of the labor market. Accordingly the author suggests, technology and automation could lead to a fairer economic model with lower prices, greater social awareness, in addition to a reduction in consumerism levels. His vision, although it might seem utopian, is that it can be already observed in dwellers of western cities where ideas related to sustainable development and economic de-growth have significantly gained relevance in the last decades.

The actual impact of the technological revolution must be contextualized within the prevailing climate of uncertainty in which the future will depend on the policies implemented in the upcoming months. On a global scale, a much more controlled globalization is expected with relevant centripetal forces, which were already visible before the pandemic (Balsa-Barreiro et al. 2020). Probably, southeastern Asia will continue to concentrate a large part of global industrial activity, but in smaller proportions than today. In the medium term, it is expected that western countries will re-shore part of their industry abroad, which will foreseeably reduce the weight of the transportation sector. On a larger scale, the survival of many companies is now under threat. However, it remains unclear which companies will be better off. It looks obvious that governments will save (and nationalize in some cases) all those companies that they define as strategic. In any case, many large companies will face great problems in surviving. Because of that, many small local businesses within a hugely complex scenario may come across new opportunities in the market. The role of state policies will determine labor costs and, therefore, the difference with regard to investment costs in automation. Regions and countries with negative demographic dynamics will only increase their productivity by optimizing industrial processes, which means more automation, or alternatively implementing a more flexible labor market with the reduction of workers' rights and the incorporation of vulnerable workers from third world countries. Therefore, factors such as the predominant size of companies, the productive model, the intervention of governments and their policies, together with the model of society that emerges after the COVID-19 pandemic will determine how disruptive the technological revolution will be and how much it will affect the labor market in the end.

Beyond the real impact of AI, automation, and robotics in the labor market, deep changes are expected to disrupt everyday life in society. Addressing the issue from the other side: without employment providing a daily structure in people's lives and with technology replacing many basic human activities, our societies will likely shift towards more individualistic entities with fewer and weaker personal interactions. For this reason, human fulfillment would play a significant role in the upcoming Industrial Revolution. After all, if technology was created to make our lives easier and more efficient in terms of time, we should wonder where the time saved goes to and even whether it is making us happier. Mayo 1947 describes that the lack of communication among people affects the social skills required to build trust and healthy communication which are fundamental for the creation of stronger social systems. Since his warnings, technology has advanced incredibly but not the capacity for people to truly communicate with each other. Moreover, challenges like facing crises and pandemics require an incredible amount of collective action which cannot be performed with a weak social system.

3 Conclusions

Before the emergence of the COVID-19 pandemic, the global labor market was already living in transition, moving towards an increasing automation because of the Tech Revolution. The sudden emergence of the pandemic has shocked the world. It will probably have several implications related to the job market, although it does not seem clear which ones yet. Drawing future scenarios is very risky at this stage. However, the current transition will have to be re-evaluated within a context of global economic crisis, which is likely to have a profound impact on the most vulnerable social classes. Today, more than ever, we must pursue sustained, inclusive, and sustainable economic growth, where the changes in the labor market are at the service of society as a whole.

Bibliography

(Arntz et al. 2016) Arntz, Melanie, Terry Gregory, and Ulrich Zierahn. 2016. The Risk of Automation for Jobs in OECD Countries: A Comparative Analysis. OECD Social, Employment and Migration Working Papers, No. 189. Paris: OECD Publishing.

(ATA 2017) American Trucking Association (ATA). 2017. Reports, Trends Statistics. Available online: https://www.trucking.org/, (accessed on 15 July 2020).

(Balsa-Barreiro 2013) Balsa-Barreiro, José. 2013. España: deuda pública, paro y crisis. Diálogo Político 2: 139-154.

(Balsa-Barreiro et al. 2020) Balsa-Barreiro, José, Aymeric Vié, Alfredo Morales, and Manuel Cebrián. 2020. Deglobalization in a hyper-connected world. Palgrave Communications 6. Available online: https://www.nature.com/articles/s41599-020-0403-x (accessed on 15 July 2020).

(Byrnes 2017) Byrnes, Nanette. 2017. As Goldman embraces automation, Even the masters of the universe are threatened. MIT Technology Review. Available online: https://www.technologyreview.com/s/603431/as-goldman-embraces-automation-even-the-masters-of-the-universe-are-threatened/ (accessed on 15 July 2020).

(Conniff 2011) Conniff, Richard. 2011. What the luddites really fought against. Smithsonian Magazine. Available online: https://www.smithsonianmag.com/history/what-the-luddites-really-fought-against-264412/ (accessed on 15 July 2020).

(Dell Technologies 2020) Dell Technologies. 2020. Accelerating digital transformation in 2020: A path forward for policymakers. Available online: https://corporate.delltechnologies.com/en-us/collaterals/unauth/white-papers/solutions/accelerating-digital-transformation-for-policymakers.pdf (accessed on 15 January 2021).

(Ding and Saenz 2020) Ding, Lei, and Julieth Saenz. 2020. Forced automation by COVID-19? Early trends from current population survey data. Discussion Papers. Federal Reserve Bank of Philadelphia. Available online: https://www.philadelphiafed.org/-/media/frbp/assets/community-development/discussion-papers/discussion-paper-automation.pdf (accessed on 15 January 2021).

(Eurostat 2020) Eurostat. 2020. COVID-19 labour effects across the income distribution. Available online: https://ec.europa.eu/eurostat/statistics-explained/ (accessed on 15 January 2021).

(Frey and Osborne, 2013) Frey, Carl Benedikt, and Michael A. Osborne. 2013. The future of employment: how susceptible are jobs to computerization? Working paper. Oxford Martin School, Oxford University (UK). Available online: https://www.oxfordmartin.ox.ac.uk/downloads/academic/future-of-employment.pdf (accessed on 15 July 2020).

(Gartner 2017) Gartner. 2017. Gartner says by 2020, Artificial Intelligence will create more jobs than it eliminates. Gartner Newsroom. Available online: https://www.gartner.com (accessed on 15 July 2020).

(Harari 2018) Harari, Yuval N. 2018. The rise of the useless class. Ideas.Ted.com. Available online: https://ideas.ted.com/the-rise-of-the-useless-class (accessed on 15 July 2020).

(Hedayatifar et al 2019) Hedayatifar, Leila, Rachel Rigg, Yaneer Bar-Yam, and Alfredo Morales. 2019. US social fragmentation at multiple scales. Journal of the Royal Society Interface 16 (159), 20190509

(Holton and Faulconbridge 2019) Holton, Kate, and Guy Faulconbridge. 2019. Thomas Cook collapses: Why and what happens now? Reuters. Available online: https://www.reuters.com/article/us-thomas-cook-grp-investment-explainer/thomas-cook-collapses-why-and-what-happens-now-idUSKBN1W804O (accessed on 15 July 2020).

(ILO 2018) International Labor Organization (ILO). 2018. Statistics and Databases. Available online: https://www.ilo.org/global/statistics-and-databases/lang–en/index.htm (accessed on 15 July 2020).

(Jaimovich and Siu 2012) Jaimovich, Nir, and Henry E. Siu. 2012. Job Polarization and Jobless Recoveries. NBER Working Paper No. w18334. Available online: https://papers.ssrn.com/sol3/papers.cfm?abstract-id=2136004 (accessed on 15 July 2020).

(Klein 2019) Klein, Christopher. 2019. The original luddites raged against the machine of the industrial revolution. History. Available online: https://www.history.com/news/industrial-revolution-luddites-workers (accessed on 15 July 2020).

(Kozmetsky and Yue 2005) Kozmetsky, George, and Piyu Yue. 2005. The Economic Transformation of the United States, 1950-2000: Focusing on the Technological Revolution, the Service Sector Expansion, and the Cultural, Ideological, and Demographic Changes. West Lafayette: Ed. Purdue University Press.

(Lu 2020) Lu, Marcus. 2020. The Front Line: Visualizing the Occupations with the Highest COVID-19 Risk. Visual Capitalist. Available online: https://www.visualcapitalist.com/the-front-line-visualizing-the-occupations-with-the-highest-covid-19-risk/ (accessed on 15 July 2020).

(Luca, Antonio, Eduardo Castelló, Yago Lizarribar, Arnaud Grignard, Luis Alonso, Dylan Sleeper, Mario Cimino, Bruno Lepri, Gigliola Vaglini, Kent Larson, and Marco Dorigo. 2019. Urban Swarms: A new approach for autonomous waste management. 2019 International Conference on Robotics and Automation (ICRA): 4233-4240.

(Manson 2015) Manson, Paul. 2015. Postcapitalism: A Guide to Our Future. London: Ed. Penguin Random House.

(Markoff 2012) Markoff, John. 2012. Skilled Work, Without the Worker. New York Times. Available online: https://www.nytimes.com/2012/08/19/business/new-wave-of-adept-robots-is-changing-global-industry.html (accessed on 15 July 2020).

(McGinty 2017) McGinty, Jo Craven. 2017. How do you fix a school-bus problem? Call MIT. Wall Street Journal. Available online: https://www.wsj.com/articles/how-do-you-fix-a-school-bus-problem-call-mit-1502456400 (accessed on 15 July 2020).

(Mayo 1947) Mayo, E., 1947. The political problem of industrial civilization. Cambridge: Harvard. University Printing Office. USA.

(Morgan 2020) Morgan, Jamie. 2020. The fourth industrial revolution could lead to a dark future. The Conversation. Available online: https://theconversation.com/the-fourth-industrial-revolution-could-lead-to-a-dark-future-125897 (accessed on 15 July 2020).

(Nedelkoska and Quintini 2018) Nedelkoska, Ljubica and Glenda Quintini. 2018. Automation, skills use and training", OECD Social, Employment and Migration Working Papers, No. 202, OECD Publishing, Paris, France. Available online: https://doi.org/10.1787/2e2f4eea-en (accessed on 15 July 2020).

(Ortega 2020) Ortega, Andrés. 2020. Coronavirus: trends and landscapes for the aftermath. Report ARI 51/2020. Real Instituto Elcano. Available online: http://www.realinstitutoelcano.org (accessed on 15 July 2020).

(Price 2020) Price, Rob. 2020. Facebook is 'opening up remote hiring aggressively' and expects half of its employees may be remote by 2030. Business Insider. Available online: https://www.businessinsider.com/facebook-half-of-all-employees-remote-2030-2020-5?IR=T (accessed on 15 July 2020).

(PwC 2017) PwC. 2017. How will automation impact jobs? Available online: https://www.pwc.com/sk/en/publikacie/the-impact-of-automation-on-jobs.html (accessed on 15 July 2020).

(PwC 2018) PwC UK. 2018. AI Will Create as Many Jobs as It Displaces by Boosting Economic Growth. Available online: https://www.pwc.co.uk/press-room/press-releases/AI-will-create-as-many-jobs-as-it-displaces-by-boosting-economic-growth.html (accessed on 15 July 2020).

(Rifkin 1995) Rifkin, Jeremy 1995. The end of work. G. P. Putnam's Sons (Schwab 2016) Schwab, Klaus. 2016. The Fourth Industrial Revolution. Cologny-Geneva: Ed. World Economic Forum.

(Semuels 2020) Semuels, Alana. 2020. Millions of Americans have lost jobs in the pandemic—and robots and AI are replacing them faster than ever—. Time. Available online: https://time.com/5876604/machines-jobs-coronavirus (accessed on 15 January 2021).

(Shendruk 2020) Shendruk, Amanda. 2020. The jobs most threatened by automation because of Covid-19. Quartz. Available online: https://qz.com/1916388/the-jobs-most-threatened-by-automation-because-of-covid-19 (accessed on 15 January 2021).

(Smith and Anderson 2014) Smith, Aaron, and Janna Anderson. 2014. AI, Robotics, and the Future of Jobs. Pew Research Center. Available online: http://www.pewinternet.org/2014/08/06/future-of-jobs/ (accessed on 15 July 2020).

(Taleb 2012) Taleb, N.N., 2012. Antifragile: Things that gain from disorder (Vol. 3). Random House Incorporated.

(Tilley 2017) Tilley, Jonathan. 2017. Automatic, robotics, and the factory of the future. MacKinsey and Company. Available online: https://www.mckinsey.com/business-functions/operations/our-insights/automation-robotics-and-the-factory-of-the-future (accessed on 15 July 2020).

(Watson 2017) Watson, Patrick W. 2017. Technology Is Already Eliminating Driver Jobs. Here's How to Trade It. Forbes. Available online: https://www.forbes.com/sites/patrickwwatson/2017/08/31/technology-is-already-eliminating-driver-jobs-heres-how-to-trade-it/215ee6253b6d (accessed on 15 July 2020).

(Weise 2019) Weise, Karen. 2019. Amazon has 30,000 open jobs. Yes, you read that right. The New York Times. Available online: https://www.nytimes.com/2019/09/09/technology/amazon-30000-job-openings.html (accessed on 15 July 2020).

(Winick 2018) Winick, Erin. 2018. Y Combinator's USD 60 million basic-income experiment will begin next year. MIT Technology Review. Available online: https://www.technologyreview.com/f/611949/y-combinators-60-million-basic-income-experiment-will-begin-next-year/ (accessed on 15 July 2020).

(World Bank 2020) World Bank. 2019. World Bank Open Data. Available online: https://data.worldbank.org (accessed on 15 July 2020).

(World Robotics 2020) International Federation of Robotics. 2020. IFR presents World Robotics Report 2020. Available online: https://ifr.org/ifr-press-releases/news/record-2.7-million-robots-work-in-factories-around-the-globe (accessed on 15 July 2020).

Health Nutrition as a Social-insurance: A complex Systems Approach in Regard to Future Pandemics

Mariana González Rodríguez M.D. and José Miguel González Rodríguez*

*Corresponding author: jm.gonzalez.rodriguez26@gmail.com

Several studies demonstrate the affect of COVID-19 virus on the development of the human body, in which the presence of metabolic disease has been linked with higher negative consequences in patients infected with the virus. Most metabolic diseases, listed as Non-Communicable Diseases (NCDs), have their roots in specific lifestyle habits of people in certain communities. This situation is an additional burden for governments across the globe who are already facing various threats to the health of their citizens including poverty, overcrowded public health systems due to previously existing diseases, underfunded state hospitals, etc. This paper highlight the need of taking a complex systems approach regarding a population's lifestyle habits, especially nutritional habits, to increase the resilience of a population when facing present and future pandemics.

1 Introduction

Lifestyle habits are a vital part of those variables, and over the last two decades, aspects related to health such as dieting, training and entertaining have been homogenized around the world (Cowen. 2002), in a so called "Western Diet," leading to a marked rise in Chronic Degenerative Diseases (CDD) (Kopp. 2019). Nutrition has become a vital component regarding health prevention in health crisis in the 21st century. Therefore, it's time to reconsider A: how we, as a society, ended up with such challenges in regards to our capacity to withstand future pandemics and B: how important of a role decision makers play in strengthening the health of the general population when facing novel stressors whether from random environmental or man-made events (Baker. 2020).

We must look at multiple factors when analyzing how to promote a healthy lifestyle around a population. Therefore, this research starts from the idea that nutrition and its subsequent challenges are complex problems from which complexity science and its findings allow decision makers to reconsider future strategies regarding events like the COVID-19 virus.

This research seeks to incentives the idea that before entering more specific ways of dealing with the "new normal" way of live, it's important to take this lesson for future generations. Namely, that the health of the population needs to be taken as an insurance policy against external events, especially for countries in Latin America that don´t have the luxury of a vast set of resources when facing these kinds of events.

2   NCDs as a systemic problem facing the COVID-19 pandemic.

Studies have shown that a high percentage of people hospitalized for COVID-19 thought to have a high mortality risk had one or more co-morbidities related to the respiratory system including diabetes, obesity, hypertension, cardiovascular disease among others. Said diseases are cataloged as Non-Communicable Diseases (NCDs) and are responsible for almost 70% deaths worldwide (WHO. 2018). They appear through time and through everyday habits of an individual, including nutritional habits, exposure to stress, the quantity of exercise, among others.

Due to the measures taken by national governments to control the outspread of the current COVID-19 pandemic, a high percentage of people around the world are staying at home. Most of them, reinforcing sedentary habits, ingest highly caloric processed food, seed oils and are being exposed to social media. The probability of developing an NCD becomes more plausible. (Porterfield, C. April 28, 2020, Creswell, J. April 07. 2020).

Because of the link between the increase of risk in patients infected with COVID-19 and the presence of NCDs, these habits represent a systemic risk to the resources of every hospital in every nation across the globe. Moreover, as more countries seek to reduce the lapse of social isolation and to "restart the economy again", even after considering the second wave of infections around the world (Parrock. May 05. 2020) , health systems with severe lack of resources may be at risk of overflowing.

Since the spread of COVID-19 worldwide, prevention measures, like social distancing, constant hand wash, mandatory use of face masks and use of antibacterial gel have been implemented, which have helped strengthen efforts for the social health of communities. However, it's important to reinforce the health of every individual in the community. As has been studied, the environment can greatly contribute to positive or negative health outcomes depending upon the structural components built of the particular society.

Just as education campaigns seek to incentives the use of face masks, it is of vital importance to reinforce healthy domestic habits during this pandemic in order to minimize the risk factors for vulnerable groups in countries where the rate of Non-Communicable Diseases is high.

We must question our current decisions regarding food consumption and production and its consequences to our health. This will lead us to reconsider the role of farm subsidies and its effects on overproduction of two specific crops, corn and soy, which are essential ingredients in most processed foods all around the world. The result has led to an homogenization of our food options, creating an illusion of a vast variety of food options, that weaken our bodies over time due to a lifetime of consumption. Also, as processed foods continued to be more affordable to the consumer relative to whole foods, most of the population around the world, especially in Latin America countries, will only be able to afford cheap foods rich in processed calories (Pollan. 2006).

3   Nutrition as a complex system.

This research parts from the idea of considering nutrition and their challenges when facing NCDs as a complex domain, characterized by being a nondeterministic and nonlinear process, where the role and their interactions between the parts in the system are extremely important. In the last decade, Complexity Science has become an important tool to better understand the challenges in the public health sector, due to the complex nature of many of the issues facing humanity regarding well-being.

In order to pass the recurrent reductionist view of the body as mechanical composition of biological factors, authors such as, Yaneer Bar-Yam *et el* (2014) advocate for a more systems-oriented solutions where the interactions and patterns between multiple variables are taken into consideration when implementing a policy strategy to NCDs.

Starting from the idea that nutrition cannot be analyzed through a single narrow perspective, especially regarding health population, Langellier *et al* (2019) present a review advocating for the use of a multilevel approach using Agent-Based Models or System-Dynamics Model. These models take into consideration how social norms shape a community, food accessibility in a specific environment, as well as the impact of household income in a specific area and preferences of individuals, and how each affect and influences nutritional needs. This provides analysts, decision makers, politicians, and the academic community with a better understanding of this topic.

The review also suggests that a complex systems approach on the topic of nutrition and health of a population can create mechanisms to better understand the network effect of patterns and feedback loops between the factors listed above. Providing a better insight when planning and understanding how diet and nutrition can be addressed and providing an information tool for local and national decision makers.

Events like the COVID-19 pandemic shows the relevance of reinforcing the battle against NCDs, due to the worsening of the infection on people with preexisting conditions. Thus, the international community needs to strengthen their efforts to achieve a better health of their populations as an insurance against external events (Guruanareobic. 2020).

It's important to understand how we as a global society, end up becoming noticeably fragile. This article cites the work of the author Michael Pollan in his book The Omnivore's Dilemma (2006), in which he explains in full detail how the idea of eating healthy is more complex as it seems. Economic, cultural, and political factors influence the way the food industry in the United States of America and other countries around the world have created a mass production of cheap ingredients to be sold globally.

Mr. Pollan states that since the 1970s, the US government has implemented a set of farm policies that incentives the overproduction of corn and soybeans domestically, which results in cheaper prices. According to the author, if the idea is to find cheap sources of energy to produce a higher quantity of food, this decision makes economic sense. As he writes, "growing corn is the most efficient way to get energy –calories- from an acre of Iowa farmland" (Ibidem. 108pp). This allows people access to a low cost and high energy food source who lack a quality source of nourishment. According to the author:

"We subsidize high-fructose corn syrup in this country, but no carrots. While the surgeon general is raising alarms over the epidemic of obesity, the president is signing farm bills designed to keep rivers of cheap corn flowing, guaranteeing that the cheapest calories in the supermarket will continue to be the unhealthiest" (Ibidem).

This is relevant because even if national efforts are made to recommend better eating habits, if the economic situation of families is not addressed, they won't be able to purchase food that will benefit the most in the long run, and will continue to consume the cheapest and highest caloric food that provide the best short-term benefit.

To prove the homogenization of food options in the market, Mr. Pollan took a mass spectrometer to calculate the atomic signature of a standard fast-food meal, to show how many of the carbon of the items came from corn and soy. The following data was gathered: "soda (100% corn), milk shake (78%), salad dressing (65%), chicken nugget (56%), and French fries (23%)" (Ibidem. 116pp).

This demonstrates that we live in a world that gives the illusion of an impressive variety of choice to the individual, However, looking through the mass spectrometer, it shows a homogenized kind of food spread all around the world.

4   Complex systems and how systems become fragile

The human body as a complex entity needs a variety of nourishment to thrive and (Taleb. 2012) the idea of feeding it food that comes from the same source, compromises it to an extent, making it less capable to resist internal or external threats. This shows how the compulsory need to optimize a specific process produce second order effect with severe consequences, that makes it difficult to identify the causes of the actual problem (Ibidem), this is important when making political and economic decisions.

The author Nassim Nicholas Taleb (2012) and the investigative journalist Gary Taubes (2007-2017) have stated that the most common dietary recommendations have been based on the first law of thermodynamics about energy conservation (known as calories in, calories out). However, when talking about a complex system, the feedback loop between the ingested food and the host plays an important role. What you consume and how you consume it and what it's made of makes a big difference in your body.

In other words, the quality of the product, how many times is consume and for how long, produce a non-linear response in the body, which weakens or strengthen the system of the host, making it less or more suited to face extreme threats by the environment or man-made events (Taleb. 2018). Just as it was shown early in the research, most of the patients who are hospitalized by the virus have a concave (negative) response when interacting with it, due to the inflammatory cascade effect of the infection.

5   A critical moment for Latin America.

Since July 2020, Latin American countries have become one of the COVID-19 hotspots, one of the highest rates of confirmed cases around the world, presenting a severe impact on the social and economic health of the region for the years to come (Horwitz, L, *et al* September 23. 2020).

In addition, the fact that one in every four people in the region live with at least one major NCDs (cardiovascular diseases like heart attacks and stroke, cancer, chronic respiratory diseases such as chronic obstructive pulmonary disease, asthma, and diabetes) affecting almost 220 million people (PAHO. June 04. 2020). This presents an important challenge to every health-system around the region, due to the danger of being overwhelmed by patients with the virus and other illnesses (France 24. August. 02. 2020).

Experts have state that "This situation in the Americas is compounded by the high prevalence of insufficient physical activity among adults, with 37.8% of women and 26.7% of men reporting insufficient physical activity (WHO 2010a; WHO 2014a). High obesity prevalence rates have been associated with higher intakes of both ultra-processed foods rich in sugars, salts, and fats and of sugar-sweetened beverages." (Ibidem. 17pp).

Consequently, in a study done by Ashktorab (et al 2020)[1] show that metabolic syndrome components like hypertension (12.1%), diabetes (8.3%) and obesity (4.5%), presented correlation with mortality across the spectrum.

Lastly, the United Nations Economic Commission for Latin America (CEPAL) in 2018 showed that most of Latin American countries only invest 4% of their GDP in healthcare (Ibidem), which presents a widespread health challenge due to the overcrowding, limited capacity and medical equipment in most state-run hospitals. Litewka and Heitman (2020) presented how this situation worsened during the pandemic because of the fact that state-run hospitals are the only choices for most families around Latin America living in poverty.

Therefore, the need to tackle the link and consequences between the COVID-19 epidemic and the NCDs to avoid a greater burden for Latin American countries is paramount (PAHO. June. 04. 2020). Efforts have been made to better address this problem including a partnership between the Pan American Health Organization and the World Health Organization to post useful information in order to tackle these challenges, focusing more into how to maintain essential services through this epidemic.

Even though plenty resources have been publish regarding NCDs and COVID-19[2], not much has been published in regards to the food industry and their impact on the health of the population, the economic disparities and the current policies incentives the overproduction of mono-cultures in order to produce cheaper ingredients for many industries around the world.

---

[1] Data collected of demographics, comorbidities and clinical symptoms from 728,282 Covid-19 positive patients in 8 Latin American countries from March 1 to July 30, 2020: Brazil, Peru, México, Argentina, Colombia, Venezuela, Ecuador and Boliva.

[2] Resources such as the "Rapid Assessment of service delivery for NCDs during the COVID-19 pandemic in the Americas" (PAHO, June. 04. 2020), "Information note on COVID-19 and noncommunicable diseases" (WHO. 2020), "THE IMPACT OF THE COVID-19 PANDEMIC ON NONCOMMUNICABLE DISEASE RESOURCES AND SERVICES: RESULTS OF A RAPID ASSESSMENT" (WHO. 2020), "Access to Essential Medicines for Noncommunicable Diseases during the COVID-19 Pandemic" (PAHO. July. 14. 2020), "Maintaining Essential Services for People Living with Noncommunicable Diseases during COVID-19" (PAHO. July. 20. 2020) and "Considerations for the Reorganization of Cancer Services during the COVID-19 Pandemic" (PAHO. May. 26. 2020)

6   Conclusion

This report highlighted the need to consider the idea of "health as an insurance against the next external event" that may threaten societies as the recurrent COVID-19 pandemic has done since early march of 2020, especially in countries in Latin America where the consequences have been devastating in several areas. Taking a complex systems approach would assist policy makers to better understand the patterns and interactions between different areas when planning actions, such as the effects of farm subsidies that allow the overproduction of crops like corn and soy all around the world, which incentives the production of cheap ingredients added to many processed foods that are widely available and accessible to the masses regardless of socioeconomic status.

In turn, this pattern results in individuals consuming the most calories at the lowest cost, creating an over-consumption of unhealthy and processed foods that weaken the body and leave them exposed when facing future external threats produce by nature or man-made events.

Bibliography

Abbany, Z. (May 14. 2020). Coronavirus: When will the second wave of infections hit?. DW News. https://www.dw.com/en/coronavirus-when-will-the-second-wave-of-infections-hit/a-53435135

Al-Maskari, F. (N,d). LIFESTYLE DISEASES: An Economic Burden on the Health Services. UN Chronicle. https://www.un.org/en/chronicle/article/lifestyle-diseases-economic-burden-health-services

Baker, M. (2020). Health as Insurance: Reducing Risk from Random Environmental and Manmade Events. 9pp

Baldwin, W., Kaneda, T., Amato, L., Nolan, L. (2013). Noncommunicable diseases in Latin America and the Caribbean: Youth are key to prevention. Washington, DC: Population Reference Bureau. https://www.prb.org/noncommunicable-diseases-latinamerica-youth-datasheet

Brito, Ch. (May 15. 2020). Coronavirus "may never go away," World Health Organization warns. CBS News. https://www.cbsnews.com/news/coronavirus-may-never-go-away-world-health-organization-endemic-virus/

Carrera-Bastos, P., Fontes-Villalba, M., O'Keefe, J. H., Lindeberg, S., Cordain, L. (2011). The western diet and lifestyle and diseases of civilization. Res Rep Clin Cardiol, 2(1), 15-35.. https://www.researchgate.net/publication/228866917

CGTN. (May 17. 2020). China's top respiratory expert Zhong Nanshan warns of potential second wave of COVID-19 infections. CGTN. https://news.cgtn.com/news/2020-05-17/Zhong-Nanshan-warns-of-potential-second-wave-of-COVID-19-infections\-QyQryYhzbi/index.html

Cowen, T. (2002). The Fate of Culture. The Wilson Quarterly. http://archive.wilsonquarterly.com/essays/fate-culture

Creswell, J. (April 07. 2020). 'I Just Need the Comfort': Processed Foods Make a Pandemic Comeback. The New York Times. https://www.nytimes.com/2020/04/07/business/coronavirus-processed-foods.html

Dehghan, M., Mente, A., Zhang, X., Swaminathan, S., Li, W., Mohan, V., Amma, L. I. (2017). Associations of fats and carbohydrate intake with cardiovascular disease and mortality in 18 countries from five continents (PURE): a prospective cohort study. The Lancet, 390(10107), 2050-2062. https://pubmed.ncbi.nlm.nih.gov/28864332/

Dellanna, L. (November, 27. 2019). The Dynamics of Risk-Taking: How Damage Affects Real-World Complex Entities, and a Framework to Estimate the Impact of Policies and Technologies on Their Survival. https://www.luca-dellanna.com/wp-content/uploads/2019/11/The-Dynamics-of-Risk-Taking.pdf

Deol, P., Evans, J. R., Dhahbi, J., Chellappa, K., Han, D. S., Spindler, S., Sladek, F. M. (2015). Soybean oil is more obesogenic and diabetogenic than coconut oil and fructose in mouse: potential role for the liver. PloS one, 10(7). https://pubmed.ncbi.nlm.nih.gov/26200659/

France 24. (August. 02. 2020). WHO warns Latin American hospitals risk being overwhelmed by Covid-19 crisis. France 24. https://www.france24.com/en/20200602-who-warns-latin-american-hospitals-risk-being-overwhelmed\-by-covid-19-crisis

Franki. R. (April 10. 2020). Almost 90 of Covid-19 Admission Involve Comorbidities. Medscape. https://www.medscape.com/viewarticle/928531

Guan, W. J., Liang, W. H., Zhao, Y., Liang, H. R., Chen, Z. S., Li, Y. M., Ou, C. Q. (2020). Comorbidity and its impact on 1590 patients with Covid-19 in China: A Nationwide Analysis. European Respiratory Journal, 55(5). https://www.medrxiv.org/content/10.1101/2020.02.25.20027664v1

Horwitz, L, et alt. (September. 23. 2020). The Coronavirus in Latin America. American Society Council of the Americas. https://www.as-coa.org/articles/coronavirus-latin-america

Infobae. (May 20. 2020). El riesgoso panorama que enfrentaría México en la segunda ola de la pandemia de coronavirus. INFOBAE. https://www.infobae.com/america/mexico/2020/05/11/influenza-y-coronavirus-combinados-el-riesgoso-panorama-que\-enfrentaria-mexico-en-la-segunda-ola-de-la-pandemia/

JIJI. (May 18. 2020). Second wave of COVID-19 cases inevitable in Japan, expert says. The Japan Times. https://www.japantimes.co.jp/news/2020/05/18/national/second-wave-covid-19-cases-inevitable-japan-expert-says/#.XsShd0RKjIU

Kopp W. (2019). How Western Diet And Lifestyle Drive The Pandemic Of Obesity And Civilization Diseases. Diabetes, metabolic syndrome and obesity: targets and therapy, 12, 2221–2236. https://doi.org/10.2147/DMSO.S216791

Legetic, B., Medici, A., Hernández-Avila, M., Alleyne, G. A., Hennis, A. (2016). Economic Dimensions of Non-Communicable Disease in Latin America and the Caribbean. Disease Control Priorities. Companion Volume. Retrieved from https://iris.paho.org/handle/10665.2/28501

PAHO. (2020). Considerations for the Reorganization of Cancer Services during the COVID-19 Pandemic. PAHO/WHO. https://iris.paho.org/handle/10665.2/52263

PAHO. (July. 14. 2020). Access to Essential Medicines for Noncommunicable Diseases during the COVID-19 Pandemic. PAHO/WHO. https://www.paho.org/en/documents/
\access-essential-medicines-noncommunicable-diseases-during-\
covid-19-pandemic

PAHO. (June. 04. 2020). Rapid Assessment of service delivery for NCDs during the COVID-19 pandemic in the Americas. PAHO/WHO. https://www.paho.org/en/documents/
\rapid-assessment-service-delivery-ncds-during-covid-19-pandemic-\
americas-4-june-2020

PAHO. (September. 20. 2020). Maintaining Essential Services for People Living with Noncommunicable Diseases during COVID-19. PAHO/WHO. https:
//iris.paho.org/handle/10665.2/52493

Parrock Jack. (May 05. 2020). Europe warned of second waves of COVID-19. euronews. mhttps://www.euronews.com/2020/05/11/
europe-warned-of-second-waves-of-covid-19

Pollan, M. (2006). The omnivore's dilemma: A natural history of four meals. Penguin Books Limited.

Porterfield, C. (April 28, 2020). In Coronavirus Quarantine, We're Eating More Processed Snacks. Forbes. https:
//www.forbes.com/sites/carlieporterfield/2020/04/28/
in-coronavirus-quarantine-were-eating-more-processed-snacks/
#67382e1d23c

Richardson, S., Hirsch, J. S., Narasimhan, M., Crawford, J. M., McGinn, T., Davidson, K. W., Cookingham, J. (2020). Presenting characteristics, comorbidities, and outcomes among 5700 patients hospitalized with COVID-19 in the New York City area. Jama. https://jamanetwork.com/journals/jama/fullarticle/2765184

Sutton, D., Harmon, N. P. (1998). Fundamentos de ecologia. Limusa. México.

Taleb, N. N. (2018, July). (Anti) Fragility and Convex Responses in Medicine. In International Conference on Complex Systems (pp. 299-325). Springer, Cham. https://link.springer.com/chapter/10.1007/978-3-319-96661-8_32

Taleb, Nassim Nicholas. Antifragile: how to live in a world we don't understand. Vol. 3. London: Allen Lane, 2012

Taubes, G. (2011). Good calories, bad calories. Anchor Books.

Taubes, G. (2011). Why we get fat and what to do about it. Anchor Books.

Taubes, G. (2017). The case against sugar. Anchor Books.

Western diet. (n.d.) Segen's Medical Dictionary. (2011). https://
medical-dictionary.thefreedictionary.com/Western+diet

WHO. (2020). The impact of the COVID-19 pandemic on noncommunicable disease resources and services: results of a rapid assessment. WHO. https://apps.who.
int/iris/handle/10665/334136

WHO. (June. 2018). Noncommunicable diseases. World Health Organization. https://www.who.int/news-room/fact-sheets/detail/
noncommunicable-diseases

WHO. (Marz. 28. 2020). Information note on COVID-19 and non-communicable diseases. PAHO/WHO. https://www.paho.org/en/documents/information-note-covid-19-and-noncommunicable-diseases

WHO. (October. 01. 2020). The best time to prevent the next pandemic is now: countries join voices for better emergency preparedness. WHO. https://www.who.int/news-room/detail/01-10-2020-the-best-time-to-prevent\-the-next-pandemic-is-now-countries-join-voices-for-better\-emergency-preparedness

Yang, J., Zheng, Y., Gou, X., Pu, K., Chen, Z., Guo, Q., ... and Zhou, Y. (2020). Prevalence of comorbidities and its effects in patients infected with SARS-CoV-2: a systematic review and meta-analysis. International Journal of Infectious Diseases, 94, 91-95. https://pubmed.ncbi.nlm.nih.gov/32173574/

Yang, Q., Zhang, Z., Gregg, E. W., Flanders, W. D., Merritt, R., and Hu, F. B. (2014). Added sugar intake and cardiovascular diseases mortality among US adults. JAMA internal medicine, 174(4), 516-524. https://pubmed.ncbi.nlm.nih.gov/24493081/

Made in the USA
Las Vegas, NV
12 May 2021